Poisonous Plants
A guide for parents & childcare providers

Elizabeth A. Dauncey

Toxicity by Leonard Hawkins and Katherine Kennedy

Kew Publishing
Royal Botanic Gardens, Kew

PLANTS PEOPLE
POSSIBILITIES

Guy's and St Thomas' **NHS**
NHS Foundation Trust
Medical Toxicology Information Services

© The Board of Trustees of the Royal Botanic Gardens, Kew and Guy's and St Thomas' NHS Foundation Trust 2010

Photographs © the photographers

The author has asserted her rights to be identified as the author of this work in accordance with the Copyright, Designs and Patents Act 1988.

All rights reserved. No part of this publication may be reproduced, stored in a retrieval system, or transmitted, in any form, or by any means, electronic, mechanical, photocopying, recording or otherwise, without written permission of the publisher unless in accordance with the provisions of the Copyright Designs and Patents Act 1988.

Great care has been taken to maintain the accuracy of the information contained in this work. However, neither the publisher nor the author can be held responsible for any consequences arising from use of the information contained herein.

First published in 2010 by
Royal Botanic Gardens, Kew
Richmond, Surrey, TW9 3AB, UK

www.kew.org

ISBN 978 1 84246 406 9

British Library Cataloguing in Publication Data
A catalogue record for this book is available from the British Library.

Production editor: Sharon Whitehead
Typesetting and page layout: Margaret Newman
Publishing, Design & Photography
Royal Botanic Gardens, Kew

Cover design: Lyn Davies

Printed and bound in Italy by Printer Trento

For information or to purchase all Kew titles please visit www.kewbooks.com or email publishing@kew.org

All proceeds go to support Kew's work in saving the world's plants for life

The paper used in this book contains wood from well-managed forests, certified in accordance with the strict environmental, social and economic standards of the Forest Stewardship Council (FSC).

Contents

Glossary 4
First aid and emergency advice from the Medical Toxicology
   Information Services 4
Introduction 5
   The plants in this book 6
   About the plant profiles 9
   In perspective 10
   How plants cause harm 11
      Poisonous if eaten (ingestion) 11
      Harmful in contact with skin 14
      Causing allergic reactions 15
      Physical injury from spines and other hazards ... 18
      Indoor plants 19
   Risk 21
A safe garden 24
   Plants and planting 24
   'Safe' plants (that children will love to grow) 25
Fruiting plants of 'low' toxicity 29
The most toxic plants (at a glance) 33
Plant profiles 34
   Outdoor plants 34
   Indoor plants 134
Sources of further information 168
Acknowledgements 169
Key to berry and flower colour 170
Index 172

Glossary

Annual – grows from seed and dies within a year

Biennial – grows from seed in the first year, flowers and then dies in the second

Deciduous – shrub or tree that loses its leaves in the autumn and regrows them in the spring

Evergreen – shrub or tree that keeps its leaves through the winter

Herbaceous – annuals, biennials and perennials that die down in the winter

Ingestion – eating

Occupational – associated with a person's employment

Perennial – lives for several years

Stamen – the part of the plant that produces pollen

Synonym (or **Syn.**) – another name by which the plant or group of plants has been known

First aid and emergency advice from the Medical Toxicology Information Services

If you suspect that someone has:

Eaten a poisonous plant:
- Do not try to make them sick.
- A glass of water or milk may be helpful.
- A spoonful of ice cream may help to relieve irritation inside the mouth.

Skin contact with sap from an irritant or allergenic plant:
- Immediately wash the affected area with warm, soapy water.
- Cover the affected area with light clothing.

Eye contact with sap:
- Rinse it immediately with clean water for 10-15 minutes if there is irritation.

If symptoms develop, or you are at all concerned, seek medical advice or attention.

Make a note of the name of the plant and if possible collect a sample, including leaves, flowers and fruit if present, to take with you.

Introduction

Laburnum and *Rhododendron* provide splashes of colour next to nettles (*Urtica dioica*)

The main purpose of this book is to provide clear and authoritative information in order to reduce the anxiety surrounding the subject of poisonous plants, and to enable you to make an assessment of the risk that plants might pose to you and those you look after. It also offers practical suggestions to help you make your home and garden safe.

This book is the latest in a series of joint publications by the Royal Botanic Gardens, Kew and the Medical Toxicology Information Services (MTIS), Guy's and St Thomas' NHS Foundation Trust.

Kew is one of the world's leading botanical institutes. As well as being the home of beautifully presented gardens that are beloved by so many people for their ornamental merit, Kew undertakes groundbreaking research on the classification and conservation of plants and fungi from all the continents of the world. The MTIS was established as Guy's Poisons Unit in 1963. The service is staffed by scientists and doctors who are experts in the field of human toxicology, and it now includes a Veterinary Poisons Information Service and a Traditional Remedies Advisory Service.

Kew and the MTIS have worked together since 1991, addressing the unique problems associated with the prevention, diagnosis and management of poisoning by plants and fungi. This collaboration has resulted in several publications including, in 2000, a CD-ROM identification system for poisonous plants and fungi, which is widely used by hospitals, local authorities and many other interested groups. For details see 'Sources of further information' on page 168.

The Horticultural Trades Association (HTA) list of potentially harmful plants, published in 2000, is another product of the collaboration between Kew and the MTIS, working with Marion Cooper and Anthony Johnson, the leading experts on British poisonous plants. Representatives of the horticultural trade, and the Royal Horticultural Society, used in-depth toxicity reports produced by the Kew–MTIS team to decide which of the ornamental plants sold in Britain should have warning labels and to agree standard phrases to describe the possible harm they might cause. The horticultural trade recognises the need to warn customers about possible risks, and the scheme is used by most major garden centres, supermarkets and DIY stores.

This book is the first publication to illustrate and describe the toxicity of all the plants on the HTA's list of potentially harmful plants and includes the HTA's risk code for each plant.

The plants in this book

This book contains 132 plants that can be harmful to humans if eaten or, in some cases, touched. Of all plants found in Britain, these are the most likely to cause harm. They include all the 117 plants on the HTA list of potentially harmful plants, and another 15 plants that are either native to Britain but not sold, such as hemlock (*Conium maculatum*), or are vegetables. Some people may be surprised to see entries for asparagus, celery, parsnip, potato and rhubarb in a book on poisonous plants, and to discover that some parts of these plants can be poisonous under certain circumstances.

To keep the book to a manageable size, it does not include harmful plants that are not commonly found in Britain, plants that are poisonous to animals but unlikely to harm humans, or fungi, which can only safely be identified by experts.

A poisonous plant border including lupin (*Lupinus*) and tobacco (*Nicotiana*)

Philodendron, usually grown as a houseplant, has been used in a tropical planting scheme with castor oil plant (*Ricinus communis*)

About the plant profiles

The main part of the book consists of profiles for individual plants. These occasionally describe a single species, such as the opium poppy (*Papaver somniferum*), but usually cover a whole group, known to botanists as a 'genus', for example *Digitalis*, which includes the foxglove (*Digitalis purpurea*), the yellow foxglove (*Digitalis grandiflora*) and other species.

The plant profiles are arranged in two sections: plants that usually grow outside, in a garden or in the countryside, and plants that are more likely to be found indoors. Because many plants that grow from bulbs can be grown both outside and indoors, particularly in the winter, all such plants have been grouped together in the outside section. You should also be aware that some plants that are usually considered to be indoor plants can be put outside in the summer in their pots or grown as tender bedding plants. These have been placed in the indoor section. Within each section similar-looking plants are grouped together as much as possible to make it easy to see the differences between them. The profile for each plant includes:

- the botanical name and most widely used common names;
- photographs, a description and symbols to show the type of plant (e.g. tree, shrub or bulb) and its location (e.g. indoors or garden). These will help you to identify the plant and to distinguish it from others that it is sometimes mistaken for. If you are not sure that you have identified the plant correctly, we recommend that you check by asking staff at a garden centre;
- brief details of the potential hazard, including the features that are most attractive to children and most likely to be involved in poisoning;
- the risk section which includes:
    - information on the number of human case enquiries that have been received by Guy's Poisons Unit. Four or less calls a year has been classified as 'Very few', 5–19 as 'Few' and 20 or more calls as 'Many'. This indicates the likelihood that a child or adult will seek medical attention for an incident relating to that particular plant;
    - the likely severity of harm that might occur from eating the plant or from skin contact;
    - the typical symptoms should the plant be eaten (ingested) and possible ill-effects from contact;
    - the HTA code indicating the potential to cause harm ('A' for plants most likely to cause harm, and 'C' for plants least likely), and a standard warning to describe the possible hazard. Although the code is voluntary, this information should be on the label of plants sold by garden centres.

In perspective

Plants are not only vital to life but they enhance our daily experience. Gardens, parks and street plantings are an amenity for us to enjoy. We live with plants all around us and they cause few cases of poisoning, **the majority are quite safe** to touch or even nibble. Judging by the enquiries received each year by poison centres, the number of children who unintentionally eat potentially poisonous plants, usually berries, is very small particularly when compared to the number who accidentally eat other poisonous substances such as prescription drugs. Most children who eat non-food plants experience no ill-effects, or, at worst, stomach ache, vomiting or diarrhoea, which quickly resolve. More serious harm is extremely rare and mostly occurs when adults intentionally eat large quantities.

Cow parsley (*Anthriscus sylvestris*) and bluebells (*Hyacinthoides non-scripta*) in late spring

How plants cause harm

Poisonous or unpleasant-tasting chemicals, as well as physical deterrents such as thorns and rough hairs, protect a plant from being eaten by animal and insect predators. Any part of a plant may contain these chemicals in various amounts: the leaves, flowers, berries, bulbs or roots. Toxicity may change according to the season of the year. As berries mature and the seeds inside become viable, they need to be eaten in order for the seeds to be dispersed. Their toxicity decreases, they lose their bitter taste and they change colour, attracting birds, rabbits, deer, badgers or even ants. When animals eat fruit the seeds are often not digested, but pass out of the body in a new location where they can germinate.

Poisonous if eaten (ingestion)

There are many different chemicals in plants that are potentially poisonous, and they can cause a range of possible harmful effects of varying severity. Some of the plants described in this book may cause immediate irritation to the mouth, others might cause stomach upset and vomiting, a few affect the heart or nervous system and can be fatal if eaten in sufficient quantity.

Seeds
Seeds can be the most toxic part of a plant. The seeds of some edible fruits are poisonous, but are seldom eaten in sufficient quantities to cause poisoning; examples include the seeds of apples and the kernels inside peach stones (see the entry for *Prunus laurocerasus* and *P. lusitanica*, page 87, which are in the same genus as peach). Seeds are not digested, so if swallowed whole and not chewed, they are unlikely to cause poisoning. Children usually spit out seeds or swallow them whole, and so may avoid toxic effects.

Some of the most commonly eaten poisonous seeds are those of *Laburnum* (see pages 12 and 119). The ripening pods resemble beans and children peel them open and eat the green seeds inside. Once the pods have dried and the seeds have become hard and black, they are less likely to be eaten in any quantity.

Flowers
Flowers are often the least toxic part of a plant, and the petals of a few species, such as nasturtiums, can even be used sparingly to add colour to a salad. However, some flowers are definitely toxic, so you should always check that you have made a correct identification before eating one. Adults have been poisoned by eating misidentified flowers. Children have occasionally been poisoned by the flowers of *Rhododendron* and related plants.

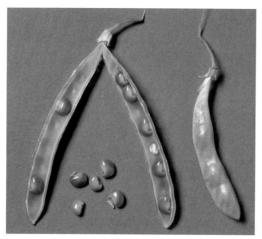

Green *Laburnum* pods look like beans, but are poisonous

The flowers of nasturtium (*Tropaeolum majus*) can be eaten

Fruit

The toxicity of fruits and berries varies considerably. Most are quite harmless to humans but a few are toxic if eaten in quantity. Some of the most commonly eaten berries are those of the widely cultivated *Cotoneaster* and *Pyracantha*, but these are of very low toxicity (see pages 29 and 32). Children are attracted to the brightly coloured, small, sweet-like fruits of these and other plants. They usually eat very few and have no or only mild ill-effects. Adults are occasionally poisoned by eating poisonous berries as food mistaking them for edible fruit. Deadly nightshade (*Atropa belladonna*) berries have sometimes been confused with blueberries.

Roots

Some roots and other underground parts, such as bulbs and tubers, can cause severe poisoning if eaten as food when mistaken for edible species. When people eat them as part of a meal, they often ignore strange tastes or the absence of the characteristic smell and eat enough to cause poisoning. Bulbs such as those of daffodils (*Narcissus*) cause abdominal pain, vomiting and diarrhoea, and the tubers of toxic members of the carrot family (Apiaceae), such as hemlock water dropwort (*Oenanthe crocata*), have caused fatalities. Children usually eat much less than an adult and experience milder symptoms.

Leaves

Poisoning by leaves is rare in Britain, but has been reported in other parts of Europe and in America. It usually occurs when poisonous species are confused with edible ones, and again it is largely adults who are at risk. Foxglove (*Digitalis*) has been confused with borage (*Borago officinalis*) and comfrey (*Symphytum*), false hellebore (*Veratrum*) has been mistaken for yellow gentian (*Gentiana lutea*), and pokeweed (*Phytolacca americana*) has been eaten instead of spinach.

Safety points

- If you suspect that a child or adult has eaten a poisonous plant, refer immediately to the 'First aid and emergency advice from the Medical Toxicology Information Services' on page 4.
- Teach children not to eat anything in the garden, parks or the countryside unless you tell them it is safe.
- Take time to identify all of the plants in your garden and then use this book to find out if they are toxic.
- If you want to gather plants from the countryside for food, use a good field guide to make sure you have identified them correctly.
- It is quite safe to compost poisonous plant material and use the resulting compost in the garden and even on your vegetable plot, but minimise direct skin contact while the plant material is still fresh.
- Store packets of seeds safely. Remember that you may have poisonous seeds in your house that have been imported from tropical countries as souvenirs such as beads, musical instruments and ornaments. Make sure these are kept where children cannot see or reach them.
- Children love to sow seeds and watch them grow, but do supervise this activity, and avoid the most toxic plants such as castor oil plant (*Ricinus communis*) and corncockle (*Agrostemma githago*).

Woody nightshade (*Solanum dulcamara*) fruits ripen from green, through orange to red

Harmful in contact with skin

Skin reactions usually arise from direct contact with a plant, particularly its sap. The areas of the body most likely to come into contact with a plant are the hands and bare arms and legs. Sap can be transferred from the hands to other parts of the body, such as the face and groin.

Plants contain chemicals that can cause various types of skin reaction:
- Irritant contact dermatitis; for example, spurge (*Euphorbia*), hellebore (*Helleborus*), hyacinth (*Hyacinthus*), and members of the Araceae, particularly dumbcane (*Dieffenbachia*);
- Phytophotodermatitis; for example, burning bush (*Dictamnus*), garden rue (*Ruta*), fig (*Ficus carica*) and some members of the Apiaceae (e.g. cow parsley (*Anthriscus sylvestris*), celery (*Apium graveolens*), hogweed (*Heracleum*) and parsnip (*Pastinaca sativa*));
- Contact urticaria, caused by stinging nettle (*Urtica dioica*);
- Allergic contact dermatitis, which is covered in the next section.

Irritant contact dermatitis usually requires prolonged or extensive exposure to the plant and is restricted to the site of exposure. The irritant can be either chemical or mechanical, or a combination of both. Mechanical irritants are stiff hairs, spines and thorns, which can penetrate the skin. Chemical irritants occur in the plant sap or in specialised cells. Reactions are not always immediate, and can vary in severity. Typical symptoms are reddening, itching, swelling and blisters that can last for several days. Any sap contact with the eye, either by transference from a hand or by squirting, can be very painful, particularly if the sap contains chemical irritants.

Phytophotodermatitis is caused by contact with the plant followed by exposure to bright UV light. Reddening and swelling of the skin can develop in 1–2 days, and blisters follow about a day later. The affected area can also be itchy and painful. The symptoms subside after several days but can leave the skin with a brown pigmentation and sensitive to sunlight for several months. Sensitive skin should be protected with high-factor sun cream or kept covered.

> *Safety points*
> - If someone gets sap on their skin or in their eye, refer immediately to the 'First aid and emergency advice from the Medical Toxicology Information Services' on page 4.
>
>   *For children*
>   - Put plants that can cause skin reactions at the back of a border where they will not easily come into contact with children playing.

- Consider keeping your garden clear of this type of plant if children play football or go onto borders for other reasons. They may brush against plants that are not normally within reach and develop skin reactions from even brief contact.
- Encourage children to wash their hands when they have been playing in the garden and before they go to the toilet to avoid transferring irritant sap from their hands to other parts of the body.
- Do not allow children to play-fight with fresh plant stalks.

For gardeners
- Take particular care when removing plants or pruning them to ensure that no harmful sap comes into contact with your skin.
- Wear protective clothing (i.e. cover all areas of skin) while strimming, and ensure that children stand well back, to avoid contact with sprayed sap. Continue to avoid skin contact when collecting up the strimmed plant material.
- Take the same precautions when pruning plants that are likely to cause skin reactions.
- Try to avoid working with plants that cause phytophotodermatitis on sunny days: even covering exposed skin may not protect you.

Causing allergic reactions

The most common types of plant-related allergic reactions are hay fever and food allergies (see 'Sources of further information' on page 168 for books and websites on these subjects). Another kind of allergic reaction that occurs on contact with plants is less well-known and affects fewer people, but can result in serious and sometimes debilitating symptoms that are often difficult to treat. The allergic reaction usually starts a few hours after exposure, but may be delayed. The skin that was in contact with the plant becomes red, and has a burning or itching sensation, swelling and possibly blistering, often in streaks. These effects may spread outwards from the point of contact and can develop over several days.

Just as everyone is not equally susceptible to hay-fever, so some people are more likely than others to develop contact allergies. Some people can be exposed on numerous occasions without developing an allergy, whereas others can become allergic after only a few exposures. In addition, some plants are more likely to cause an allergy (i.e. are more allergenic) than others. Poison ivy (*Rhus radicans*), rarely

Ivy (*Hedera helix*) growing over a fence

Peruvian lily (*Alstroemeria*) flowers can be a variety of colours

grown in Britain but widespread in North America, can cause very serious allergic reactions on only the second contact. Children are less likely to be allergic to plants than adults as they will have handled plants less.

Direct contact is not always necessary to induce a reaction. Susceptible individuals can develop itching eyes, a runny nose and a non-productive cough from simply being in the same room as a plant that they are allergic to. When pruning or shredding garden plants such as ivy (*Hedera*), air-borne allergens can cause a rash on exposed skin (e.g. on the face) without direct plant contact.

Allergic reactions are most commonly reported in people who handle plants in the course of their work (occupational exposure). Examples include bulb growers and packers, particularly those working with tulip (*Tulipa*), daffodil (*Narcissus*) and hyacinth (*Hyacinthus*), and florists and floral designers, who often strip the leaves from the lower stems of plants such as Peruvian lily (*Alstroemeria*). Symptoms are usually restricted to the finger-tips and other areas of the hands. Cold and wet work conditions can cause the skin of the hands to crack and make the symptoms worse. Horticulturists and gardeners, particularly those who grow a lot of one type of plant, such as *Chrysanthemum*, can also have sufficient exposure to become allergic.

The main causes of occupational plant contact dermatitis amongst horticulturists, bulb and flower packers, and florists are included in this book, and a few of the plants causing less commonly reported problems are mentioned here. These are usually safe to grow in a domestic setting where they will not be handled repeatedly or for prolonged periods. *Chrysanthemum* (page 133) and several other members of the daisy family (Asteraceae) can cause allergic reactions in sensitised individuals. These include *Achillea* (yarrow), *Cichorium* (chicory), *Cynara scolymus* (globe artichoke), *Dahlia*, *Helenium* (sneezeweed), *Helianthus annuus* (sunflower), *Bracteantha* (syn. *Helichrysum*, straw flower), and *Tanacetum* (feverfew, tansy). Irritant or allergic reactions have also been reported for *Allium* (onion), *Anigozanthos* (kangaroo paws), *Asclepias curassavica* (blood flower), *Codiaeum* (variegated croton), *Coleus*, *Hydrangea*, *Laurus nobilis* (bay, laurel), *Anemone patens* (pasque flower), the orchids *Cymbidium*, *Cypripedium* and *Paphiopedilum*, and the ferns *Arachnioides adiantiformis* (leather leaf fern) and *Nephrolepis exaltata* (sword fern, ladder fern, Boston fern).

Unfortunately this list is not exhaustive. Although this book includes the plants that most frequently cause allergic reactions in Britain or cause the most severe effects, many other plants affect a few individuals or are problems in other parts of the world. You are referred to 'Sources of further information' on page 168 if you would like to know more about the subject.

> *Safety points*
> - If your child is susceptible to allergies, for example if they have eczema, you may want to exclude allergenic plants from your house and garden.
> - If you become allergic to a plant, avoid further contact and consider excluding the plant from your home and office.
> - The allergens have been bred out of particular strains of one allergenic plant, German primula (*Primula obconica*), so check the labels (or, to avoid contact, ask someone to check them for you).
> - Cover up when pruning or shredding allergenic garden plants.
> - Clothes should be carefully washed after contact with these plants to prevent further spread of the allergen.

Physical injury from spines and other hazards

Several common garden shrubs have thorns or spines on their stems, and sometimes also on their leaves. Gardeners have been encouraged to grow such plants (e.g. *Berberis*, *Pyracantha* or holly (*Ilex*)) along boundaries and under windows to deter burglars. Others plants that can cause physical injury have ornamental value (e.g. roses) or are grown for their soft fruits (e.g. brambles, raspberries and gooseberries).

Thorns and spines are not only harmful in the obvious way, through direct damage to the skin. They can also introduce bacteria into the wound, which becomes inflamed and painful as a result. In addition, the tips of spines and thorns can break off inside the wound, and if they remain there they can cause irritation resulting in the formation of a granuloma (a lump) that requires surgical removal.

Some grasses and sedges have sharp edges to their leaves; the leaves of pampas grass (*Cortaderia*) are particularly tough and have serrated leaf edges that can cut and irritate skin. The stems of hops (*Humulus lupulus*) have rows of stiff bristles that catch the skin and leave scratches. The tips of *Yucca*, *Agave* and some *Aloe* leaves are very hard and pointed, and have caused eye injuries when people have bent over them.

Fremontodendron is a shrub or small tree that is often trained against a wall to give it both support and some winter protection. The young stems and backs of the leaves are covered with a coating of fine hairs that easily detach, particularly during pruning. The hairs become airborne and can irritate the eyes, nose, respiratory tract and skin. They can also be transferred from hands to the face or other parts of the body.

Pot-grown *Agave parryi* has sharp leaf tips   Spiny *Pyracantha*, firethorn

Safety points
- Avoid growing thorny or spiny plants near paths, lawns or other areas used by children.
- When pruning, wear protective clothing including heavy-duty gardening gloves.
- Collect up all prunings so that there are no thorny stems left on the ground where they might be trodden on.
- It is not advisable to shred or compost prunings at home as the spines will remain intact and may cause harm when used in the garden later, e.g. as a mulch. Instead, consider bagging up prunings and taking them to the council green waste site.

Indoor plants

Cacti
Many cacti have sharp spines that can cause painful skin pricks. Some are hooked and easily attach to the skin and clothing. Of particular note are the short, barbed hairs (glochids) of *Opuntia* species, including bunny ears (*Opuntia microdasys*), which are grown as house plants.

Pot-grown bulbs
Spring can be brought into the house early with pots of bulbs. Children will enjoy planting them and watching the shoots grow. The most toxic part of these plants is the bulb, with leaves, stems and flowers being less of a hazard.

Araceae
Many members of the arum family, the Araceae, are popular houseplants. They include the peace lily (*Spathiphyllum*), painter's palette (*Anthurium*) and *Philodendron*. They all contain calcium oxalate crystals, which cause immediate pain and irritation to the mouth and throat if any part of the plant is chewed. The most notorious member of this group, the dumbcane (*Dieffenbachia*), probably contains the highest concentrations of calcium oxalate and is likely to cause the most severe symptoms.

Poinsettia
Unlike many other species of *Euphorbia* (see pages 99 and 151), poinsettia (*Euphorbia pulcherrima*) contains only low levels of chemical irritants. Because it is thought to be poisonous, numerous cases of poinsettia ingestion have been reported to poisons centres around the world, yet in most instances there were no symptoms and only a few people experienced mild irritation of the mouth and stomach.

Cut flowers
The potential toxicity of cut flowers is usually overlooked when they are bought and sold. A recent, and most welcome, exception is the warning on pre-packaged lilies that they are harmful to cats. This book includes a symbol for 'cut flower' in the list of locations, so you will know which plants to look out for at the florists. The most toxic cut flower is probably monkshood (*Aconitum*) (see page 125).

Opuntia microdasys, bunny ears, and a variety of other cacti and succulents

Lily pollen is particularly toxic to cats

Safety points
- Keep all indoor plants and cut flowers out of the reach of curious children.
- Supervise children when planting bulbs; sensitive individuals should wear gloves.
- Dispose of any water that cut flowers have stood in; poisoning has occurred when this water has been drunk.

Risk

When assessing the risk associated with a plant we must consider:
- the toxicity of the plant and its various parts;
- the person who might eat or touch it;
- the type of plant or the part of the plant they are most likely to eat or touch;
- how much of the plant they might eat.

When considering the person who might eat or touch a plant, their age is probably the most important factor. Infants from 12 to 36 months of age develop excessive hand to mouth activity and are attracted to bright colours. They frequently put things in their mouths but eat very little. Up to the age of 5 or 6, children are still at risk because they are curious but have no sense of danger. They compare things to sweets and food such as peas and so might eat several berries or seeds, but this is rarely enough to cause anything but mild symptoms. When adolescents and adults eat plants, however, they do so intentionally. They think the plant is edible and eat it as food, they use it for recreation for its psychological effects, or they intend to harm themselves. In nearly all instances, they are likely to eat amounts large enough to contain significant amounts of poison if the plant is toxic.

To take an example, the leaves and seeds of the yew tree (*Taxus*) are very poisonous, but the red 'berries' (arils) surrounding the seed are harmless. The leaves are tough so unlikely to be eaten. The red, sweet-tasting arils are attractive to children and it is quite likely that a curious child might pick and eat one. However, because the seed is hard, it is seldom, if ever, chewed or even swallowed. As a result, although poison centres have received numerous reports of children possibly eating yew berries, very few of them were poisoned. There is a significant risk of a child eating the fruit if it is accessible, but a very low risk that they will be poisoned as result. The risk is not great enough to justify removing yew trees from gardens where young children play, but it would be sensible to consider surrounding it with a fence to keep children (and dogs) away from fruit on the ground.

Basal rosette of foxglove (*Digitalis purpurea*) leaves

As another example, the risk of an individual being poisoned by foxglove (*Digitalis*) can be assessed as follows. Foxgloves are a common wild flower and are also frequently grown in gardens. All parts of the plant are poisonous if eaten, with the flowers containing the lowest concentration of toxin. Foxgloves have bright flowers but the fruits are green capsules rather than berries, so it is unlikely to be attractive to small children. They may play with the flowers but rarely eat them. The most likely circumstance in which poisoning could occur is if adults ingest the plant mistaking it for something else. The leaves have a bitter taste that would deter an adult from eating them raw, but they have been used to make herbal tea by people who have mistakenly identified it as another plant, such as comfrey (*Symphytum*). Tea made from foxglove leaves is likely to contain a significant amount of poison and someone could drink a poisonous amount fairly easily. However, few people make herbal tea from leaves that they collect themselves and the likelihood that they will mistakenly collect foxglove leaves instead of those of a non-toxic plant is small. The risk of an individual being poisoned by foxglove is therefore extremely small even though it is a poisonous plant.

First identify the plants in your garden; photograph of a poisonous plants show garden 'The Dark Side of Beauty' (see p. 169)

A safe garden

With some practical advice, children can enjoy plants safely. Whether you want to grow just a few plants to add visual variety to your back garden or are keen to introduce your children to growing fruit and vegetables, a few simple steps can give you peace of mind when the children are unsupervised.

The first step is to identify the plants already in the garden. If they are poisonous, you will find them in the following pages; if they are not included in this book they are probably of low risk.

Poisonous plants do not necessarily have to be removed from your garden. There are a few steps that you can take to manage the potential hazard to babies, children and yourself. Below you will find some ideas on how you can design a safer garden, that is full of plants. Earlier sections of this book have included 'Safety points' which give you advice on how to minimise the risk from particular types of poisonous plant, and how to handle them safely during routine gardening activities. For advice about general garden safety see 'Sources of further information' on page 168.

Plants and planting

- Grow ornamental plants and edible fruit, vegetables and herbs in separate areas of the garden.
- When creating a vegetable patch, be sure to dig out and remove all ornamental bulbs from the area.
- Trees that bear tempting pods or other fruit can have the lower branches pruned away so that the remaining fruits are inaccessible, see image on page 23, which shows pruned *Robinia* trees.
- If you do not want to spoil the shape of the tree, then keep young children away with a physical barrier such as a deep, well-planted flower bed or even a fence.
- If you are choosing a tree for your garden, try a cultivar that does not fruit, such as *Laburnum* × *watereri* 'Vossii' or male cultivars of yew (*Taxus*).
- Fruiting hedges, such as cherry laurel (*Prunus laurocerasus*) and privet (*Ligustrum*), should be pruned directly after flowering so that fruits do not develop.
- Other fruiting herbaceous plants or shrubs can have the flowers removed as they die to prevent them from fruiting.
- Poisonous herbaceous plants can be transplanted to the back of the border.

'Safe' plants (that children will love to grow)

As with the plants in the 'low' toxicity section (p. 29), the plants listed below are not edible, except for the vegetables and herbs. No symptoms are expected if small amounts are eaten, or if the plants are handled during normal gardening activities. A few exceptions are included as a note. Irritation of the mouth and mild stomach upset are possible if larger amounts are eaten, and some people may experience a rare allergic reaction to any of these plants. Most of the following plants will be available from garden centres but some can only be grown from seed. Always wash hands after handling seeds as they may have been dressed with fungicides, etc. before packing.

P = perennial   B = biennial   A = annual

Touch

Herbaceous

FALSE DITTANY (*Ballota pseudodictamnus*) (P) soft, hairy leaf texture

ELEPHANT'S EARS (*Bergenia*) (P) smooth, leathery leaves; red-leaved varieties are prettier

CORAL BELLS (*Heuchera*) (P) rough, crinkly leaves

Potentilla atrosanguinea (P) soft, hairy leaf texture

PLANTAIN LILY (*Hosta*) (P) ribbed, textured leaves

SILVER SAGE (*Salvia argentea*) (B) silvery, soft leaves

LAMB'S EARS (*Stachys byzantina*, syn. *S. lanata*) (P) soft, hairy leaf texture

GIANT SILVER MULLEIN (*Verbascum bombyciferum*) (B) silvery, soft leaves

Grasses

QUAKING GRASS (*Briza maxima*) (A) seed heads rustle in the wind

HARE'S TAIL (*Lagurus ovatus*) (A) soft, fluffy seed heads

FOUNTAIN GRASS (*Pennisetum alopecuroides*) (P) soft, fluffy seed heads

PONYTAILS (*Stipa tenuissima*) (P) soft, fluffy seed heads

Ferns

LADY FERN (*Athyrium*) (P) feathery foliage

HART'S TONGUE FERN (*Asplenium scolopendrium*) (P) shiny, leathery foliage

HARDY SHIELD FERN (*Polystichum aculeatum*) (P) shiny, leathery foliage

Flowers

LOVE LIES BLEEDING (*Amaranthus caudatus*) (A) long tassels of furry flowers; other species have edible leaves and seeds

GRANNY'S PINCUSHION (*Astrantia*) (P) papery flowers

Ageratum (A) soft, fluffy flowers

ORNAMENTAL ONIONS (*Allium*) (P) softly, spiky flowers

SNAPDRAGON (*Antirrhinum*) (A) petals open like a mouth when gently squeezed at the side

POMPOM DAISY (*Bellis perennis*) (B) tactile, pompom flowers

STRAW FLOWER (*Bracteantha bracteata*, syn. *Helichrysum bracteatum*) (A) stiff straw-like 'everlasting' flowers; keep at back of border to restrict contact

INDIAN SHOT PLANT (*Canna indica*) (P) tender; leaves have a ridged texture

Dahlia (P) tender; pompom types are very tactile; an example of the Fibonacci sequence

Fuchsia (P) some tender; double flowers very tactile

BABY'S BREATH (*Gypsophila*) (P) small, dense flowers

GAYFEATHER (*Liatris spicata*) (P) spikes of fluffy flowers

STATICE (*Limonium sinuatum*) (A) papery, 'everlasting', brightly coloured flowers

LOVE IN A MIST (*Nigella damascena*) (A) dried seed heads rattle

Zinnia (A) pompom types are very tactile

Scent

Scented flowers or leaves will attract bees and other insects.

Herbaceous

SWEET SULTAN (*Amberboa moschata*, syn. *Centaurea moschata*) (A) fluffy flowers

ORNAMENTAL HYSSOP (*Agastache*) (P) leaves have a range of scents

CHOCOLATE COSMOS (*Cosmos atrosanguineus*) (P) tender; chocolate scent

PINKS (*Dianthus*) (P) clove-like scent; petals can be eaten; carnations rarely have scent

SWEET WILLIAM (*Dianthus barbatus*) (A)

HELIOTROPE (*Heliotropium arborescens*) (A) also has fuzzy leaves

MIGNONETTE (*Reseda odorata*) (A)

LAVENDER (*Lavandula*) (P)

BEE BALM, OSWEGO TEA (*Monarda didyma*) (P) aromatic leaves

EVENING PRIMROSE (*Oenothera biennis*) (B)

Phlox paniculata (P)

BROMPTON STOCK (*Matthiola incana*) (A)/(B)

SWEET PEA (*Lathyrus odoratus*) (A) choose a scented variety; remove dead flowers to maintain flowering and stop the inedible seed pods forming

ORNAMENTAL SAGE (*Salvia*) (P) leaves have a range of fruity scents but some do smell unpleasant

Scabiosa atropurpurea (A) soft, pincushion-type flowers; loved by butterflies

VIOLET (*Viola odorata*) (P)

WALLFLOWER (*Erysimum* and *Cheiranthus*) (B)/(P)

Shrubs
- BUTTERFLY BUSH (*Buddleja*) some varieties have deep purple or white flowers; butterflies are particularly attracted to the flowers
- ROSE (*Rosa*) choose strongly scented flowers; pick thornless varieties for safety
- LILAC (*Syringa*)
- SHRUBBY HONEYSUCKLE (*Lonicera fragrantissima*) flowers in winter
- *Viburnum*; some have scented flowers in winter, others fruit
- WINTERSWEET (*Chimonanthus praecox*) flowers in winter

Climbers
- HONEYSUCKLE (*Lonicera periclymenum*) (P)

Herbs

These herbs are considered safe when used in the small quantities normally used in cooking.
- BASIL (A) varieties come in a range of scents and flavours
- CHIVES (P)
- FENNEL (A) hoverflies are attracted to the seed heads
- MARJORAM (A) some have golden leaves
- MINT (P) varieties come in a range of scents and flavours
- ROSEMARY (P)
- SAGE (P)
- THYME (P) varieties come in a range of scents and flavours

Vegetables
- CABBAGE (A)
- SWISS CHARD or RUBY CHARD (*Beta vulgaris*) (A) leaves and midribs have been bred in a range of bright colours; use young leaves raw in salads but older ones are better cooked
- LETTUCE (A) some varieties have frilly and coloured leaves
- PEA (A)
- STRAWBERRY (P) alpine strawberries have small fruits throughout the summer

Colour

Bright colours but with no particular scent or touch interest.

Summer bedding
- GRANNY'S BONNET (*Aquilegia*) (P)
- *Arabis* (P)
- *Aubretia* (P)
- ENGLISH MARIGOLD or POT MARIGOLD (*Calendula officinalis*) (A) petals can be eaten
- CANTERBURY BELLS (*Campanula medium*) (B)

COCKSCOMB (*Celosia*) (A)

CORNFLOWER (*Centaurea cyanus*) (A)

Cosmos bipinnatus (A)

CALIFORNIA POPPY (*Eschscholzia californica*) (A)

Gazania (P) tender

SUNFLOWER (*Helianthus annuus*) (A) an example of the Fibonacci sequence

BUSY LIZZIE (*Impatiens walleriana*) (A)

POACHED EGG PLANT (*Limnanthes douglasii*) (A) very attractive to hoverflies

BEDDING LOBELIA (*Lobelia erinus*) (A) not the herbaceous species (see page 123)

MONKEY FLOWER (*Mimulus*) (A)

Nemesia strumosa (A)

Osteospermum ecklonis (P) tender

GERANIUM (*Pelargonium*) (P) tender; some have nicely scented foliage

POLYANTHUS (*Primula* Polyanthus Group) (P) bright and cheerful in early spring

NASTURTIUM (*Tropaeolum majus*) (A) flowers and seeds are edible

Verbena speciosa (A)

PANSY (*Viola* × *wittrockiana*) (P) some flowers have cheeky faces

Herbaceous

MONTBRETIA (*Crocosmia masonorum*) (P)

CRANESBILL, HARDY GERANIUM (*Geranium*) (P)

Penstemon (P)

Scabiosa (P)

Verbena bonariensis (P) butterflies are particularly attracted to the flowers

Veronica (P)

Bulbs

Chionodoxa (P)

Crocus vernus (P) spring flowering only

Iris e.g. *reticulata* (P) spring flowering only

GRAPE HYACINTH (*Muscari*) (P) *Muscari latifolium* is more attractive with two-tone blue flowers

Fruiting plants of 'low' toxicity

Many fruiting plants that are commonly eaten by children are considered to be non-toxic or to have only a low toxicity. In most cases there will be no symptoms. Occasionally, the child may experience irritation of the mouth and mild stomach upset. Some of these plants are illustrated here to assist with their identification.

Aucuba japonica, spotted laurel, with red fruit. The leaves of other forms of *Aucuba* are completely green or with splashes of yellow.

Flowering and fruiting *Berberis darwinii*. Many species of *Berberis* have spiny leaves.

A red leaved form of *Berberis thunbergii*.

Cotoneaster horizontalis has small leaves. Other species of *Cotoneaster* can form large shrubs or small trees. The fruit can be single or in clusters, and are red, orange or yellow.

Hawthorn, *Crataegus monogyna*, is a small tree with lobed leaves and clusters of red fruit.

False castor oil plant, *Fatsia japonica*, has large lobed leaves like those of the castor oil plant (see p. 98). Its clusters of black fruit are similar to those of ivy (see p. 91).

Some species of *Fuchsia* set fruit.

The fruit of tutsan, *Hypericum androsaemum*, are quite dry unlike the juicy fruit of deadly nightshade (see p. 71).

Hypericum × *inodorum* 'Magical Red'.

Holly, *Ilex*, often has spiny leaves. The fruit are usually red but can be orange or yellow.

The leaves of true laurel, also known as bay, *Laurus nobilis*, are aromatic when crushed.

The climbing honeysuckle, *Lonicera periclymenum*, forms clusters of sticky red berries.

Lonicera pileata is a low-growing shrub that can produce pairs of bright purple fruit.

The Duke of Argyll's tea plant, *Lycium barbarum*, occasionally produces red fruit. The dried fruit of that species, and *L. chinense*, shown here, are sold as goji berries.

The sweet-smelling yellow flowers of *Mahonia japonica* are followed by fruit that ripen to blue-purple.

The Chinese lantern, *Physalis alkekengi*, is named for the papery structures that enclose the fruit

An orange-fruited form of firethorn, *Pyracantha*.

It is more unusual for *Pyracantha* to have yellow fruit.

The mountain ash, or rowan, *Sorbus aucuparia*, is a small tree with large clusters of orange fruit.

The blueberry, *Vaccinium corymbosum*, is cultivated for its edible fruit.

Other species of *Sorbus* can have yellow or pink fruit, and some, such as *S. hupehensis*, even have white fruit.

The most-toxic plants (at a glance)

Abrus precatorius (lucky beans, rosary pea) p. 154
Aconitum (monkshood, wolf's bane) p. 125
Arum (cuckoo pint, lords-and-ladies) p. 35
Atropa belladonna (deadly nightshade) p. 71
Brugmansia (angels' trumpets, tree datura) p. 140
Cicuta virosa (cowbane, water hemlock) p. 105
Colchicum (autumn crocus, naked ladies) p. 62
Conium maculatum (hemlock) p. 103
Convallaria (lily-of-the-valley) p. 43
Coriaria (tutu) p. 83
Daphne (mezereon, spurge laurel) p. 86
Datura (jimsonweed, thornapple) p. 68
Dieffenbachia (dumbcane, leopard lily) p. 156
Digitalis (foxglove) p. 126
Euphorbia (spurge) pp. 99, 151
Gloriosa superba (flame lily, glory lily) p. 144
Heracleum (hogweed) p. 106
Hyoscyamus (henbane) p. 67
+*Laburnocytisus* 'Adamii' (Adam's laburnum) p. 118
Laburnum (golden rain) p. 119
Lantana (shrub verbena) p. 85
Mandragora (mandrake) p. 69
Nerium oleander (oleander) p. 115
Oenanthe crocata (hemlock water dropwort) p. 104
Phytolacca (inkberry, pokeweed) p. 84
Primula obconica (German primula, poison primula) p. 138
Rhus radicans (poison ivy) p. 92
Ricinus communis (castor oil plant) p. 98
Ruta (garden rue) p. 111
Scopolia (Russian belladonna) p. 70
Solandra (chalice vine) p. 142
Solanum dulcamara (woody nightshade) p. 75
Sophora (kowhai) p. 117
Taxus (yew) p. 82
Thevetia (yellow oleander) p. 147
Veratrum (false hellebore) p. 52
Zigadenus (death camas) p. 53

Outdoor plants

Plants usually found growing outside

ARUM
cuckoo pint, Italian lords-and-ladies, lords-and-ladies

Family
Araceae

Description
Short plants at the base of hedges and other partly shaded places, also grown in gardens. Large, sometimes patterned, leaves and unusual 'flowers' are frequently followed in summer by heads of orange to scarlet berries (3–14 mm). The baked and ground tubers of *Arum maculatum* were once used like arrowroot.

Main toxin
Calcium oxalate

Risk
Many reported cases. Ingestion is most likely to result in mild to moderate poisoning.

Symptoms
Ingestion of the plant or its sap can cause salivation, burning sensation in the mouth and sore throat, swelling of the lips, nausea, vomiting and abdominal pain. Drowsiness and dizziness may occur.

Contact with the sap may cause irritation of the skin or eye.

HTA Category
B, CAUTION toxic if eaten; skin and eye irritant

Flowering plants of *Arum italicum*, showing leaves

Fruiting plants of *Arum maculatum* growing through ivy

ARISAEMA

cobra-lily, Jack-in-the-pulpit, snail flower

Family
Araceae

Description
Short to medium garden plants, preferring partial shade. Striking, lobed leaves and unusual 'flowers', consisting of a spathe that can be hooded, striped and with an elongated tip, appear in the spring or summer. Heads of fruit (6–12 mm) ripen to orange or red.

Main toxin
Calcium oxalate

Risk
Very few reported cases. Ingestion or contact may result in mild poisoning.

Symptoms
Ingestion may result in immediate burning of the mouth and gastrointestinal tract, often with difficulty in swallowing and sometimes with vomiting and diarrhoea.

Contact with the sap may result in immediate irritation of the skin or eye.

HTA Category
C, Harmful if eaten; skin and eye irritant

Arisaema ciliatum var. *liubaense*, flowering

Arisaema tortuosum, fruiting

DRACUNCULUS
dragon arum

Family
Araceae

Description
Medium garden plants preferring full sun or some shade. In spring, they produce large, foul-smelling 'flowers', formed from a purple spathe and a long, central, almost black spadix, which attract flies. Sometimes followed by a head of orange fruit (8–15 mm).

Main toxin
Calcium oxalate

Risk
Very few reported cases. Ingestion or contact may result in mild poisoning.

Symptoms
Ingestion is likely to cause immediate irritation in the mouth and a burning sensation in the throat. Mild gastrointestinal upset with nausea and vomiting may occur.

Contact may result in mild to moderate irritation of the skin and eye.

HTA Category
C, Harmful if eaten; skin and eye irritant

The flat, purple spathe mimics raw meat

A head of ripe fruit

Tall plants of *Dracunculus vulgaris* with unopened flowers

ZANTEDESCHIA
arum lily, calla lily, florists' calla

Synonym
Calla hort.

Family
Araceae

Description
House, conservatory or garden plants requiring damp conditions. Bearing lush leaves and erect, funnel-shaped 'flowers', which may be white or brightly coloured yellow, pink or nearly black. Sometimes followed by heads of yellow fruit (8–15 mm).

Main toxin
Calcium oxalate

Risk
Very few reported cases. Ingestion is most likely to result in mild poisoning.

Symptoms
Ingestion of the plant or its sap can cause a burning sensation in the mouth and throat; other clinical effects are unlikely to occur.

Contact with skin and eye can cause local irritation and inflammation.

HTA Category
C, Harmful if eaten; skin and eye irritant

Zantedeschia aethiopica, ripe fruit head (© RBG, Kew)

Zantedeschia aethiopica, flowering plants

Zantedeschia elliottiana with yellow spathes

CALLA PALUSTRIS
bog arum, water arum

Family
Araceae

Description
Growing in very shallow water at the edge of ponds or rivers in gardens, and occasionally in the countryside. A creeping mat of bright green heart-shaped leaves. Small 'flowers' with a white spathe may be followed by heads of fruit (5–8 mm) that ripen to a dull red.

Main toxin
Calcium oxalate

Risk
Very few reported cases. Ingestion or contact may result in mild poisoning.

Symptoms
Ingestion may cause immediate stinging, irritation and blistering of mucous membranes in the mouth and throat, resulting in salivation, swelling and difficulty in swallowing.

Contact may result in mild dermatitis. Eye exposure is likely to cause irritation.

HTA Category
C, Harmful if eaten; skin and eye irritant

Notes
Calla is a common name for *Zantedeschia*.

Close-up of *Calla palustris* 'flower' (© Sonny Larsson)

Fruiting plant (© Waterside Nursery)

LYSICHITON
skunk cabbage

Family
Araceae

Description
Growing at the edge of water in gardens, and occasionally in the countryside. Large, yellow or white 'flowers' grow in the spring just before or at the same time as the leaves. The large, simple leaves continue to grow after the 'flowers' have died down.

Main toxin
Calcium oxalate

Risk
Very few reported cases. Ingestion or contact may result in mild poisoning.

Symptoms
Ingestion or chewing of plant material is expected to cause immediate stinging, irritation and blistering of mucous membranes in the mouth and throat, resulting in salivation and difficulty in swallowing.

Contact may result in mild dermatitis. Eye contact with the sap is likely to cause irritation.

HTA Category
C, Harmful if eaten; skin and eye irritant

Lysichiton americanus (© Mark Jackson)   Lysichiton camtschatensis (© Mark Jackson)

ACTAEA section ACTAEA
baneberry, black cohosh, herb Christopher

Family
Ranunculaceae

Description
Shade-loving garden plants with divided leaves; *Actaea spicata* is a rare British plant. Heads of small, white flowers in late spring are followed in summer by attractive red, white or black berries (5–15 mm). A herbal preparation is made from *A. racemosa*.

Main toxin
Uncertain

Risk
Very few reported cases. Ingestion of small quantities may result in moderate poisoning.

Symptoms
Ingestion of all parts, especially roots and fruits, causes an intense pain in the mouth and throat followed by vomiting, diarrhoea and abdominal pain. In severe cases there may be dizziness, disturbed vision or perception and convulsions. Kidney damage may occur.

Contact with some species can cause skin irritation and blisters.

HTA Category
C, Harmful if eaten; skin irritant

White fruit of *Actaea pachypoda*

Red fruit of *Actaea rubra* ssp. *arguta*

Actaea rubra f. *neglecta*, flowering

IRIS
gladdon, yellow flag

Family
Iridaceae

Description
A large group of wild and garden plants, growing in sun or shade, dry or wet conditions. Showy flowers, with parts in 3s, are produced from early spring to summer. Dry fruit capsules usually contain more or less dry, flattened seeds, but the seeds of *Iris foetidissima* are rounded (6–8 mm) and bright reddish-orange.

Main toxin
Unknown irritant

Risk
Few reported cases. Ingestion may result in mild poisoning.

Symptoms
Ingestion can cause a burning sensation in the mouth and throat, followed by abdominal pain, nausea, vomiting and diarrhoea.

Contact can cause irritation and rash.

HTA Category
C, Harmful if eaten

Iris foetidissima, flower and bud, also showing the sword-like leaves

Iris foetidissima, fruit capsules splitting to reveal the orange seeds (© Mark Jackson)

CONVALLARIA
lily-of-the-valley

Family
Ruscaceae (syn. Convallariaceae)

Description
Small native plants, also widely cultivated, spreading in open woodland and other partly shaded places. Slender heads of fragrant, white or occasionally pink, nodding, bell-shaped flowers grow with 2 or 3 simple leaves. Sometimes followed by bitter-tasting, orange or red berries (8–10 mm).

Main toxins
Cardiac glycosides

Risk
Very few reported cases. Ingestion of large quantities may result in severe poisoning.

Symptoms
Ingestion may cause gastrointestinal irritation, nausea and vomiting. A large quantity may result in a reduced heart rate and blood pressure.

Occupational contact may cause dermatitis.

HTA Category
B, CAUTION toxic if eaten

Convallaria majalis, flowering plants showing paired leaves   *Convallaria majalis*, fruiting plants

POLYGONATUM
Solomon's seal

Family
Ruscaceae (syn. Convallariaceae)

Description
Shade-loving, wild and garden plants. Leaves simple, alternating or in whorls. Greenish-white, nodding, bell-like flowers appear in summer and are followed by blue-black or red berries (4–10 mm).

Main toxins
Saponins and possibly cardiac glycosides

Risk
Very few reported cases. Ingestion may result in mild poisoning.

Symptoms
Ingestion may result in nausea, vomiting and diarrhoea. Theoretically, there may also be visual disturbance, disorientation and convulsions. Irregular heart rhythm and low blood pressure are possible but extremely unlikely.

HTA Category
C, Harmful if eaten

Polygonatum maximowiczii with blue-black fruit

Polygonatum latifolium, flowering

Polygonatum verticillatum with red fruit

ASPARAGUS
asparagus fern, climbing asparagus, foxtail fern

Family
Asparagaceae

Description
The young shoots (asparagus tips) of *Asparagus officinalis* are eaten as a vegetable. Also found in the wild, and some species are houseplants. Finely branched stems with much reduced leaves and leaf-like branches. Small white or cream flowers can be followed by red berries (5–11 mm).

Main toxins
Possibly saponins

Risk
Very few reported cases. Ingestion or contact may result in mild poisoning.

Symptoms
Ingestion of fruit may result in gastrointestinal upset with vomiting, abdominal pain and diarrhoea. Localised oral allergic symptoms may occur in individuals sensitive to *Asparagus*.

Contact may result in dermatitis and allergic contact urticaria with redness of the skin and itching. In sensitive individuals, eye contact may cause itching, conjunctivitis and swelling of the eyelids and inhalation can result in runny nose, asthma, tight throat, hoarseness, sneezing and coughing.

HTA Category
C, May cause skin allergy; fruits harmful if eaten (as *Asparagus* except *A. officinalis*)

Asparagus officinalis, fruiting

Uncut asparagus flowers in the summer

The houseplant *Asparagus densiflorus* occasionally fruits
(© RBG, Kew)

BRYONIA DIOICA
red bryony, white bryony

Synonym
Bryonia cretica subsp. *dioica*

Family
Cucurbitaceae

Description
A hedgerow plant that climbs using corkscrew tendrils. Roughly hairy, lobed leaves. Flowers greenish, from spring to summer, with female plants producing clusters of small, red fruits (5–10 mm) in the autumn.

Main toxins
Triterpenes called cucurbitacins

Risk
Very few reported cases. Ingestion of small quantities may result in mild poisoning.

Symptoms
Ingestion may result in repeated vomiting, abdominal pain and diarrhoea (which may be bloody). Irritation of the mouth and nose, dizziness and significant difficulty in breathing have been reported.

Contact with fresh tubers may result in redness of the skin, painful inflammation and blistering.

HTA Category
n/a

Flowering plant scrambling over a fence

Fruiting plant

Fruit can persist after the leaves have died

DIOSCOREA COMMUNIS
black bryony

Synonym
Tamus communis

Family
Dioscoreaceae

Description
Climbing wild plants found scrambling through hedges and over fences. Stems grow each year from large black tubers. Leaves heart-shaped. Flowers small, greenish-yellow, with female flowers followed in late summer and autumn by clusters of shiny red fruits (10–15 mm) that persist after the leaves have fallen.

Main toxins
Calcium oxalate and irritant compounds

Risk
Very few reported cases. Ingestion and contact are most likely to result in mild poisoning.

Symptoms
Ingestion can cause local irritation and inflammation of the mouth and throat, which may be severe. Nausea and diarrhoea are also possible.

Contact can cause dermatitis. Sap rubbed on the skin can result in burning, reddening, painful swelling and a rash. Allergic reactions are also possible.

HTA Category
n/a

Male flowering plant with heart-shaped leaves

Female plant with ripening fruit, growing through hawthorn

IPOMOEA PURPUREA and I. TRICOLOR
morning glory

Synonyms
Ipomoea rubrocaerulea, *I. violacea* hort.

Family
Convolvulaceae

Description
Tender plants grown in conservatories or in gardens in the summer. Climbing by stems winding around a support. Leaves simple, heart-shaped. Showy, trumpet-shaped, short-lived flowers open in the morning and close later in the day. Sometimes followed by dry capsules containing grey to black seeds (3.5–5 mm).

Main toxins
Indole alkaloids

Risk
Very few reported cases. Accidental ingestion may result in mild poisoning. Intentional ingestion may result in moderate poisoning.

Symptoms
Ingestion may result in restlessness, nausea, abdominal pain, vomiting, diarrhoea, facial flushing, blurred vision, drowsiness, disorientation, numbness of the extremities and muscle tightness. There may be disturbances in hearing and vision, which can be distressing. Individual response is variable. Large ingestions may also result in shock and low blood pressure.

HTA Category
C, Harmful if eaten

Ipomoea tricolor 'Heavenly Blue'

Ipomoea purpurea 'Star of Yelta'

Seed capsules of *Ipomoea purpurea* (© RBG, Kew)

PASSIFLORA
passion flower, passion fruit

Family
Passifloraceae

Description
Climbing plants grown in gardens or conservatories. Leaves deeply divided, sometimes simple. Unusual, showy flowers have numerous petals, a ring of filaments and prominent stamens. Sometimes producing large, orange fruit (35–60 mm; seeds 4–7 mm). The edible fruits of some tender species are sold in Britain.

Main toxins
Possibly harman alkaloids and some species contain cyanogenic glycosides

Risk
Few reported cases. Severe poisoning is unlikely.

Symptoms
Ingestion may cause gastrointestinal upset. There is a possibility of cyanide poisoning but this is rare.

HTA Category
C, Harmful if eaten (as *Passiflora caerulea*)

Ripe fruit broken open to reveal fleshy pink seeds (© RBG, Kew)

Passiflora caerulea, flowering plant

Passiflora caerulea, fruiting plant (© Mark Jackson)

PODOPHYLLUM
May apple

Family
Berberidaceae

Description
Low-growing, shade-loving, garden plants with 1 or 2 large, variously lobed leaves that are sometimes patterned when young. Flowers, produced in the spring, are usually solitary and white, sometimes more than one and dark red. Fruits large (25–50 mm), red or yellow.

Main toxins
Lignans, including podophyllotoxin

Risk
Very few reported cases. Inappropriate use by adults can result in severe poisoning.

Symptoms
Ingestion can cause vomiting, diarrhoea and abdominal pain, which can be followed by lethargy, drowsiness and lack of coordination. Severe poisonings have been reported following misuse of *Podophyllum* resin (a herbal preparation).

HTA Category
C, Harmful if eaten

Young *Podophyllum hexandrum* leaves

Podophyllum peltatum, flowering

Podophyllum hexandrum, fruiting in the summer

HELLEBORUS
bear's foot, Christmas rose, hellebore

Family
Ranunculaceae

Description
Occasionally found in the wild, but widely grown in gardens. Short or robust plants, with leathery leaves, divided like a hand into leaflets or lobes. Nodding flowers in late winter and spring. Segmented fruit capsules split open when dry.

Main toxins
Cardiac glycosides and protoanemonin, the amounts vary between species and different parts of the plant

Risk
Very few reported cases. Ingestion of small quantities may result in severe poisoning. Prolonged contact may cause moderate effects.

Symptoms
Ingestion may result in a burning sensation, blistering and ulceration in the mouth, vomiting, diarrhoea and abdominal pain. Disorientation, convulsions and cardiac symptoms are possible.

Contact with fresh plant material may cause skin irritation and a sensation of burning within a few minutes. Prolonged contact (upwards of an hour) may cause redness, swelling and blistering of the skin, with symptoms developing over about 36 hours. Nasal irritation and severe irritation of the eye may also occur.

HTA Category
C, Harmful if eaten; skin irritant

A *Helleborus* cultivar flowering in the spring.

Green fruit capsules of *Helleborus foetidus*, one split open to show the seeds (© RBG, Kew).

VERATRUM
false hellebore

Family
Melanthiaceae

Description
Shade-loving garden plants. Leaves have a pleated appearance. Branched flowerheads bear numerous white, green or dark flowers.

Main toxins
Veratrum alkaloids

Risk
Very few reported cases. Ingestion may result in severe poisoning. Contact may result in mild poisoning.

Symptoms
Ingestion may result in a burning pain in the throat, nausea, profuse vomiting, diarrhoea and abdominal pain. Other possible effects include sweating, thirst, low blood pressure and pulse, shock, collapse, coma and irregular heart rhythm. Neurological effects are less common and may include dizziness, paraesthesia, confusion and convulsions.

Contact can result in mild skin irritation. Eye exposure may result in severe irritation, eye watering, pain and inflammation.

HTA Category
B, CAUTION toxic if eaten

Flowering plant of *Veratrum album* (© Robert Bevan-Jones)

Young leaves of *Veratrum nigrum*

Part of a flower-head of *Veratrum nigrum*

ZIGADENUS
death camas

Synonym
Zygadenus

Family
Melanthiaceae

Description
Shade-loving plants, occasionally grown in gardens or with some winter protection. Leaves linear, sometimes keeled. Large heads of greenish- or yellowish-white, star-shaped flowers are borne on long stems in late spring and early summer.

Main toxins
Veratrum alkaloids

Risk
Very few reported cases. Ingestion may result in severe poisoning.

Symptoms
Ingestion of bulbs may result in nausea, vomiting, abdominal cramps and diarrhoea, as well as in decreases in pulse and blood pressure, lack of coordination and muscle twitching. Cardiovascular symptoms may be fatal or resolve within 24 hours, but gastrointestinal symptoms may persist for 48 hours.

Contact can result in skin irritation.

HTA Category
B, CAUTION toxic if eaten

Zigadenus venenosus, young flowerhead and leaves

Zigadenus venenosus, flowering

ORNITHOGALUM
chincherinchee, star-of-Bethlehem, wonder flower

Family
Hyacinthaceae

Description
Found in the wild, particularly in grassland, also grown in gardens and greenhouses and sold as cut flowers. Leaves narrowly to broadly linear. In spring and summer, white, or occasionally yellow or orange, star-shaped flowers are produced in large heads, often as the leaves are dying.

Main toxins
Cardiac glycosides and calcium oxalate

Risk
Very few reported cases. Ingestion of large quantities may result in severe poisoning.

Symptoms
Ingestion can cause nausea and vomiting. Ingestion of a large quantity may result in an irregular heart rhythm and decrease in blood pressure.

Contact with the sap of some species, e.g. *O. longibracteatum*, is irritant and can cause stinging and reddening of the skin.

HTA Category
C, Harmful if eaten

Cut flowers of *Ornithogalum saundersiae*

O. umbellatum has a green stripe on the outside of the petals

SCILLA
autumn squill, spring squill

Family
Hyacinthaceae

Description
Occasionally found in grassland, but more commonly grown in gardens, particularly the spring-flowering *Scilla siberica*. Heads of blue or occasionally pink, purple or white flowers, which are nodding or upright, bell-shaped, flat or star-shaped, are produced from spring to autumn.

Main toxins
Piperidine alkaloids and possibly cardiac glycosides

Risk
Very few reported cases. Ingestion may result in severe poisoning.

Symptoms
Ingestion will usually result in gastrointestinal upset only. Ingestion of a very large quantity may result in oral pain, nausea, vomiting, abdominal pain and diarrhoea.

HTA Category
C, Harmful if eaten

Notes
Some species of *Scilla* are now in other genera such as *Hyacinthoides* (page 56). Squill is also used for other plants not in this book.

Scilla siberica, flowering plant with bulb (© RBG, Kew)

Scilla peruviana, flowering

S. siberica, flowers and immature fruit capsules (© RBG, Kew)

HYACINTHOIDES
bluebell, Spanish bluebell

Synonyms
Endymion, *Scilla campanulata*, *S. italica*, *S. non-scripta* and *S. nutans*

Family
Hyacinthaceae

Description
Found in woods (often carpeting the ground in blue in spring) and hedgerows, also widely cultivated. The flowers are usually blue but can be white or pink, more-or-less nodding, narrow to broadly bell-shaped, on straight or curving long stems. Fruits (10–15 mm; seeds, 2–4 mm).

Main toxins
Piperidine alkaloids and possibly cardiac glycosides

Risk
Few reported cases. Ingestion of a large amount may result in severe poisoning.

Symptoms
Ingestion will usually result in gastrointestinal upset only. Ingestion of a very large quantity may result in oral pain, nausea, vomiting, abdominal pain and diarrhoea.

HTA Category
C, Harmful if eaten

Bluebells (*Hyacinthoides non-scripta*) with nodding heads of slender flowers and narrow leaves

Spanish bluebells (*Hyacinthoides hispanica*) with upright heads of wide flowers and broad leaves

HYACINTHUS
hyacinth

Family
Hyacinthaceae

Description
Grown in gardens and parks for their sping-time flowers, and also indoors in pots for an earlier display. Fragrant flowers are produced in dense heads in a variety of colours: blue, purple, pink, orange, yellow and white. Heads have fewer flowers in their second and following years.

Main toxins
Piperidine alkaloids and calcium oxalate

Risk
Few reported cases. Ingestion or contact can result in mild to moderate poisoning.

Symptoms
Ingestion may cause rapid onset of vomiting, diarrhoea and abdominal discomfort.

Occupational contact with bulbs can lead to 'hyacinth itch', where fingertips become cracked and reddened. This can sometimes progress to an eczematous dermatitis, which may involve the hands, arms and face.

HTA Category
C, Harmful if eaten; skin irritant

Naturalised flowering bulbs

Blue and white flowering bulbs with spring bedding

Bulb cut in half to show arrangement of scales (© RBG, Kew)

NARCISSUS

daffodil, jonquil, Lent lily

Family
Amaryllidaceae

Description
Occasionally found in the wild; widely grown in gardens and parks for their spring flowers; also grown in pots indoors for a Christmas display. Flowers are usually single but sometimes in small heads, yellow, cream, white, sometimes with orange, with a distinctive central trumpet.

Main toxins
Amaryllidaceae alkaloids, calcium oxalate crystals and allergens

Risk
Many reported cases. Ingestion or contact may result in mild to moderate poisoning.

Symptoms
Ingestion can rapidly lead to severe diarrhoea, vomiting and abdominal pain. Less frequently, there may be light-headedness, salivation and shivering.

Contact with sap can lead to a rash on the forearms and hands, which can develop into eczema. It may spread all over the body, but typically affects the arms, neck, face and thighs. Handling the bulb can result in severe dermatitis, particularly of the fingers. Those most at risk are growers, florists and those who pack the flowers and bulbs.

HTA Category
C, Harmful if eaten; skin irritant

Flowering plants in light woodland

A mass of flowering plants (© Mark Jackson)

Bulb cut in half to show arrangement of scales (© RBG, Kew)

TULIPA
tulip

Family
Liliaceae

Description
Bulbs widely grown in the garden and cool greenhouse. Small to large flowers open in late spring and early summer, usually bowl-shaped but can open flat or have finely drawn out tips to the petals. Flowers yellow, orange, red, mauve, purple, pink or white, single coloured or marbled.

Main toxins
Allergens (tulipalin A and B)

Risk
Few reported cases. Ingestion or contact may result in mild to moderate poisoning.

Symptoms
Ingestion of bulbs may result in nausea, vomiting, abdominal pain and diarrhoea. Sweating, salivation, difficulty in breathing and palpitations have also been reported.

Occupational contact frequently results in allergic contact dermatitis. There may be painful tingling and redness of the fingertips, particularly around the fingernails, within 24 hours of handling bulbs. Secondary spread to the arms, face and genitalia can occur. Eye exposure may result in conjunctivitis and swelling of the eyelids. Rhinitis, sneezing, wheezing and cough have been reported from inhalation.

HTA Category
C, Harmful if eaten; may cause skin allergy

Flowering plants in a variety of colours

Striped flowers underplanted with wallflowers

Bulbs, some cut to show scales (© RBG, Kew)

HIPPEASTRUM
amaryllis

Family
Amaryllidaceae

Description
Large bulbs, usually grown indoors to flower in the winter but can be put outside in the summer. Flowers are produced a few together on long stems, very showy, in white, pink or red, or occasionally cream or yellow, sometimes striped or edged in another colour.

Main toxins
Amaryllidaceae alkaloids

Risk
Very few reported cases. Ingestion or contact may result in mild poisoning.

Symptoms
Ingestion may result in nausea, vomiting and diarrhoea. Effects are expected to occur rapidly and to persist for 3–4 hours. Light-headedness, shivering and salivation are possible.

Contact may result in dermatitis.

HTA Category
C, Harmful if eaten

Large red flowers and strap-shaped leaves

A dry bulb (© RBG, Kew)

AMARYLLIS BELLADONNA
belladonna lily, Jersey lily

Family
Amaryllidaceae

Description
Grown in sheltered spots in gardens or in cool greenhouses. In the autumn, long stems bear a cluster of somewhat large, light to dark pink or white, trumpet-shaped flowers. The leaves are produced after flowering.

Main toxins
Amaryllidaceae alkaloids

Risk
Very few reported cases. Ingestion is most likely to result in mild poisoning.

Symptoms
Ingestion can lead to nausea, vomiting and diarrhoea, possibly with dizziness and drowsiness.

HTA Category
C, Harmful if eaten

Notes
Amaryllis is the common name for *Hippeastrum*.

Close-up of flowers

Long flowering stems appear before the leaves

Leaves grow after the flowers

COLCHICUM
autumn crocus, meadow saffron, naked ladies

Synonyms
Bulbocodium, Merendera

Family
Colchicaceae

Description
Uncommon in the wild in damp meadows and woods, but widely cultivated. Flowers resemble a large crocus, are purple, pink or white and produced in the autumn. Leaves grow in the spring and are somewhat large and broadly strap-shaped. Seed capsules rattle when dry.

Main toxins
Colchicine alkaloids

Risk
Very few reported cases. Ingestion may result in severe poisoning. Contact may result in mild poisoning.

Symptoms
Ingestion may result in immediate irritation of the mouth. After a delay of 2–12 hours, common effects include nausea, vomiting, diarrhoea, abdominal pain, increased heart rate and chest pain. More severe effects such as low blood pressure, low heart rate, convulsions, irregular heart rhythm and death have been reported.

Contact may result in skin irritation.

HTA Category
B, CAUTION toxic if eaten

A cultivated *Colchicum* with chequered petals

Colchicum speciosum seed capsules

Dry corms of *Colchicum bornmuelleri*

MORAEA section HOMERIA
Cape tulip

Synonym
Homeria

Family
Iridaceae

Description
Unusual plants growing from corms (similar to bulbs), in cool greenhouses or conservatories. Leaves narrowly linear or strap-shaped. Flowers open in succession on a long stem, often scented, quite showy in orange, peach or yellow.

Main toxins
Cardiac glycosides

Risk
Very few reported cases. Ingestion may result in severe poisoning.

Symptoms
Ingestion can result in severe nausea and vomiting, as well as in weakness, dilated pupils, heart problems and collapse.

HTA Category
C, Harmful if eaten (as *Homeria*)

Moraea longistyla flowers

Moraea elegans flowers

Moraea collina has very narrow leaves

ALSTROEMERIA
lily of the Incas, Peruvian lily, St Martin's flower

Family
Alstroemeriaceae

Description
Grown in gardens and sold as long-lasting cut flowers. Tall stems are leafy throughout their length. Loose heads of flowers are produced in summer. Flowers are spreading trumpet-shaped, in yellow, orange, pink, red, or white, the inner 3 petals are often patterned with yellow patches and dark spots.

Main toxin
An allergen (tulipalin A)

Risk
Very few reported cases. Contact may result in mild to moderate allergic reactions.

Symptoms
Ingestion has not been reported, but mild gastrointestinal upset is possible if this plant is eaten.

Occupational contact commonly causes allergic contact dermatitis. There may be reddening of the skin, eczema, itching, fissuring and scaling of the fingers and hands, and secondary spread to the arms, face and genitalia.

HTA Category
C, May cause skin allergy

Alstroemeria 'Tessa'

Alstroemeria aurea

MIRABILIS
marvel of Peru, four o'clock plant, false jalap

Family
Nyctaginaceae

Description
Slightly tender garden plants with long tap-roots. Stems branching, leafy throughout, leaves oval, broadest at the base. Flowers slightly fragrant, opening in the evening and closing the next morning, brightly coloured pink, magenta, yellow or white, sometimes marbled. Seeds large (7–9 mm) and black.

Main toxin
Unknown

Risk
Very few reported cases. Ingestion or contact may result in mild poisoning.

Symptoms
Ingestion may cause stomach pain, nausea, vomiting, abdominal cramps and diarrhoea.

Contact may result in mild skin irritation.

HTA Category
C, Harmful if eaten; skin irritant

Bright pink is a common colour for *Mirabilis jalapa* flowers

Yellow-flowered plant

Black seeds form in the cup of the calyx

NICOTIANA
tobacco

Family
Solanaceae

Description
Tender garden plants, usually grown as summer bedding. Leaves sometimes sticky and pungent-smelling. Flowers often fragrant in the evening, and can be purple, pink, cream, white or green.

Main toxins
Alkaloids (nicotine)

Risk
Many reports of exposure to tobacco products. Accidental ingestion by children and inappropriate use by adults can result in severe poisoning. Repeated contact can result in mild effects.

Symptoms
Ingestion is followed within 2 hours by nausea, vomiting, salivation and abdominal pain. Diarrhoea occurs occasionally and may be delayed by 4–24 hours. There may be initial increases in heart rate, breathing and blood pressure, followed by sudden decreases in these signs if large amounts are ingested or absorbed. Pupils are initially constricted and then later dilated. In severe cases, there may also be confusion, agitation and restlessness followed by coma, convulsions and cardiac effects.

Contact may result in redness of the skin, urticaria, mild eczema and allergic contact dermatitis, usually as a result of repeated exposure.

HTA Category
C, Harmful if eaten

Nicotiana sanderae 'Malibu Blue' and 'Malibu Lime'

Nicotiana sylvestris has nodding, white flowers

HYOSCYAMUS
henbane

Family
Solanaceae

Description
Rare wild plant found particularly near the coast, also occasionally grown in gardens. Sticky, strong smelling, softly hairy plants. Leaves somewhat large, jagged toothed. Flowering in the summer, flowers cream or yellow, often veined with purple. Fruit a capsule.

Main toxins
Tropane alkaloids

Risk
Very few reported cases. Ingestion of small quantities may result in moderate poisoning.

Symptoms
Ingestion can cause a dry mouth, blurred vision, warm and dry skin and dilated pupils. Other clinical effects include drowsiness, confusion, increased heart rate, reduced bowel sounds, difficulty in passing urine and disturbed vision or perception.

Contact can result in blistering and dermatitis. Contact with the eye can cause dilated pupils.

HTA Category
B, CAUTION toxic if eaten

Hyoscyamus niger, curved flowering stem from the side, with immature fruit capsules

The flowers of *Hyoscyamus niger* sometimes lack the purple veining

DATURA
Jamestown weed, jimsonweed, stinkweed, thornapple

Family
Solanaceae

Description
Annual weed of cultivated ground, also grown as an ornamental. Large, pungent-smelling leaves, either simple or with jagged teeth. Flowers large, trumpet-shaped, occasionally double, sweetly fragrant, white, cream or mauve. Fruit an egg-shaped, often spiny, capsule (25–70 mm; seeds 2–4 mm).

Main toxins
Tropane alkaloids

Risk
Very few reported cases. Ingestion may result in severe poisoning.

Symptoms
Ingestion can cause dry mouth, blurred vision, dilated pupils, reduced bowel sounds, difficulty in passing urine and disturbed vision or perception. Increased heart rate and flushed face may also be present.

Eye contact with plant material or sap may cause dilated pupil(s).

HTA Category
B, CAUTION toxic if eaten

Notes
See also the tender, perennial *Brugmansia* (page 140).

Double-flowered cultivar

Datura 'La Fleur Lilac' with mature seed capsule splitting open to show black seeds

MANDRAGORA
mandrake

Family
Solanaceae

Description
Rarely cultivated plant with a long, branching tap-root. A rosette of large leaves grows before or with the flowers. The violet to greenish-white flowers appear in autumn to spring. Large green fruit (30–60 mm), ripening to yellow or orange, occasionally follow in the summer.

Main toxins
Tropane alkaloids

Risk
Very few reported cases. Ingestion may result in moderate poisoning.

Symptoms
Ingestion can cause a dry mouth, blurred vision, warm and dry skin and dilated pupils. Other clinical effects include confusion, increased heart rate, reduced bowel sounds, difficulty in passing urine and disturbed vision or perception.

Eye contact with plant material or sap may cause dilated pupil(s).

HTA Category
B, CAUTION toxic if eaten

M. officinarum with flowers and leaves emerging in spring

Mandragora officinarum with smooth leaves and immature fruit

M. officinarum can have purple flowers and rough leaves

SCOPOLIA
Russian belladonna

Family
Solanaceae

Description
Rarely cultivated, shade-tolerant plants that die back after flowering. The short, fleshy stems grow in late spring and are leafy throughout. Flowers are produced singly where a leaf joins the stem, dull yellow or reddish to purple with yellow inside, bell-shaped.

Main toxins
Tropane alkaloids

Risk
Very few reported cases. Ingestion may result in severe poisoning.

Symptoms
Ingestion can cause dry mouth, blurred vision, dilated pupils, increased heart rate, warm dry skin, reduced bowel sounds, difficulty in passing urine and disturbed vision or perception.

HTA Category
B, CAUTION toxic if eaten

Flowering plants of *Scopolia carniolica* in spring

ATROPA BELLADONNA
belladonna, deadly nightshade, dwale

Family
Solanaceae

Description
Large, somewhat rare, native plants, also occasionally grown in gardens or occurring as weeds. Leaves simple. Purplish, bell-shaped flowers in early summer are followed by juicy, black berries (10–20 mm), with conspicuous leaf-like parts (calyx) at the base. Superficially similar to tutsan, *Hypericum androsaemum* (page 30).

Main toxins
Tropane alkaloids

Risk
Many reported cases. Ingestion may result in severe poisoning.

Symptoms
Ingestion can cause dry mouth, blurred vision and dilated pupils, increased heart rate, warm and dry skin, reduced bowel sounds, difficulty in passing urine and disturbed vision or perception.

Contact may result in dermatitis. Dilated pupil(s) may occur if plant material or sap touches the eye.

HTA Category
B, CAUTION toxic if eaten

Flowering plant   Close-up of fruiting branch

SOLANUM NIGRUM
black nightshade

Family
Solanaceae

Description
Weed of waste and cultivated ground, appearing in mid-summer. Small, white flowers with prominent yellow stamens are followed by hanging clusters of small fruit (6–10 mm), that ripen to black. Fruit are occasionally found as a contaminant of frozen peas and beans.

Main toxins
Steroidal alkaloids, absent from completely ripe berries

Risk
Very few reported cases. Ingestion is most likely to result in mild poisoning.

Symptoms
Ingestion can cause mild gastrointestinal upset. More severe symptoms are possible but rarely occur.

HTA Category
n/a

Flowering plant with clusters of unripe, green fruit

Close-up of plant with unripe green and ripe black fruit

SOLANUM TUBEROSUM
potato

Family
Solanaceae

Description
Widely grown root vegetable. Large plants with divided leaves, the leaflets varying in size. The white or mauve flowers are sometimes followed by green, tomato-like fruits (15–40 mm). The tubers (potatoes) turn green when exposed to sunlight, pink-skinned cultivars darken.

Main toxins
Steroidal alkaloids

Risk
Very few reported cases. Ingestion is most likely to result in mild poisoning.

Symptoms
Ingestion of fruit or inappropriately stored, green or sprouting potatoes can cause gastroenteritis, which may be delayed. More severe symptoms, including high temperature, confusion, low blood pressure and difficulty in breathing, have been reported but rarely occur.

Contact can occasionally cause an allergic dermatitis.

HTA Category
n/a

Flowering plants

A single, green fruit

Exposed potatoes

SOLANUM (other cultivated ornamental species)
Chilean potato tree, kangaroo apple, nightshade, potato vine

Family
Solanaceae

Description
Cultivated shrubs, scrambling or climbing plants. Leaves simple or variously toothed or lobed. Bearing blue, purple or white flowers with a central point of yellow stamens. Sometimes producing fruit (6–100 mm), that ripen to yellow, orange, red, black or occasionally purple.

Main toxins
Steroidal alkaloids

Risk
Few reported cases. Ingestion is most likely to result in mild poisoning.

Symptoms
Ingestion can cause mild nausea, vomiting and diarrhoea. More severe effects are uncommon.

Contact with some species may result in skin irritation.

HTA Category
C, Harmful if eaten (cultivated ornamental species except *S. dulcamara*)

Notes
This group includes *Solanum aviculare*, *S. crispum*, *S. laxum*, *S. laciniatum* and *S. rantonettii*.

Solanum crispum 'Glasnevin', a climbing plant with purple flowers

Solanum laciniatum, a large, shrubby plant with fruit that ripen to orange

Solanum laxum 'Album' (syn. *S. jasminoides* 'Album'), a scrambling climber, with fruit that ripen to black

SOLANUM DULCAMARA
bittersweet, woody nightshade

Family
Solanaceae

Description
Common climbing or sprawling plants of hedges, woodland, waste ground and gardens; a variegated form is sold as a garden plant. Leaves are simple, or with 1–4 lobes at the base. Loose clusters of bright purple flowers are followed by berries (6–15 mm), that ripen in the summer through orange to red (also see page 13).

Main toxins
Steroidal alkaloids, ripe fruit contain only small amounts

Risk
Few reported cases. Ingestion may result in moderate poisoning.

Symptoms
Ingestion may cause an intensely bitter taste in the mouth, which may be followed by a sweet aftertaste (hence the name bittersweet). Most cases have no symptoms, but nausea, vomiting, diarrhoea and drowsiness are occasionally reported. Rarely, there may be irritation of the mouth and throat, headache, thirst, dizziness, weakness, fever, increased heart rate and difficulty in breathing.

HTA Category
B, CAUTION toxic if eaten

Flowering plant, the leaves have 2 lobes

Fruiting plant, with the fruit at various stages of ripeness

A coastal variety with simple leaves

SOLANUM PSEUDOCAPSICUM
Christmas cherry, Jerusalem cherry, winter cherry

Synonyms
Solanum capsicastrum, S. diflorum

Family
Solanaceae

Description
Small shrubs grown as house or garden plants. Small white flowers in early summer are followed by green fruits (9–25 mm), that ripen to orange or red, occurring mainly from autumn through the winter, but may persist into the summer.

Main toxins
Steroidal alkaloids

Risk
Few reported cases. Ingestion may result in mild poisoning.

Symptoms
Ingestion may result in mild gastrointestinal upset, possibly abdominal pain, vomiting and diarrhoea. More serious effects are extremely unusual.

HIA Category
C, Harmful if eaten

Bushy plant with fruits and flowers

Various parts of a plant (© RBG, Kew)

CAPSICUM ANNUUM
chilli pepper, chillies, ornamental pepper

Family
Solanaceae

Description
Edible sweet and chilli peppers are grown under glass or outside in the summer; fruiting plants are also sold as ornamentals in the autumn. Small to medium plants with simple leaves. Small flowers are followed by attractive, elongated fruits (20–150 mm), that can be yellow, orange, red or purple.

Main toxin
Capsaicin

Risk
Few reported cases. Ingestion or contact may result in moderate poisoning.

Symptoms
Ingestion may cause a burning sensation in the mouth, throat and stomach. There may be diarrhoea, abdominal cramping, difficulty in passing urine or faeces, inflammation of the stomach, abdominal distension and vomiting. Gastrointestinal perforation has been reported.

Contact may result in prolonged burning pain, irritation and redness of the skin without blistering. Allergic reactions, from dermal and oral exposure have been reported in latex-sensitive individuals. Eye exposure may result in burning pain, eye watering and corneal abrasions. Inhalation may cause transient breathing difficulties.

HTA Category
C, Skin and eye irritant; harmful if eaten (as *Capsicum annuum* (ornamental cultivars))

Chilli pepper 'Numex Twilight', with flowers and young fruits

Ornamental chilli peppers in a variety of colours

GAULTHERIA MUCRONATA
pernettya, prickly heath, wintergreen

Synonym
Pernettya

Family
Ericaceae

Description
Small to large, evergreen shrubs grown in gardens, and very occasionally found in the countryside. Leaves very small, stiff, with a sharp tip. Flowers small, urn-shaped, in summer. Female plants produce white, lilac, pink, red or deep purple fruits (5–17 mm), that persist into winter.

Main toxins
Grayanotoxins

Risk
Very few reported cases. Ingestion of large quantities may result in severe poisoning.

Symptoms
Ingestion of a few fruits may cause gastrointestinal upset. Following large ingestions, initial effects could include vomiting, a burning and/or numb sensation in the mouth, salivation and abdominal pain. Other possible symptoms are slow heart rate, abnormally low blood pressure, irregular heart rhythm and neurological effects.

HTA Category
C, Harmful if eaten (as *Gaultheria* (section *Pernettya* only))

Notes
This account does not include *Gaultheria procumbens* from which we get 'oil of wintergreen'.

A deep-pink fruiting cultivar

White-fruited cultivar (© Mark Jackson)

SYMPHORICARPOS
snowberry

Family
Caprifoliaceae

Description
Shrubs grown in gardens and found elsewhere, particularly in damp places such as riverbanks. Slender, arching stems with widely spaced, simple, rarely lobed leaves. Tiny pink, bell-shaped flowers are followed by white or pink, puffed berries (6–20 mm).

Main toxin
Unknown

Risk
Few reported cases. Ingestion of small quantities is most likely to result in mild poisoning.

Symptoms
Ingestion can sometimes cause nausea, vomiting and diarrhoea, and possibly dizziness and drowsiness.

HTA Category
C, Harmful if eaten

Symphoricarpos albus, fruiting and flowering plant

Fruiting branches of *Symphoricarpos* × *chenaultii* 'Hancock'

VISCUM ALBUM
mistletoe

Family
Santalaceae (syn. Viscaceae)

Description
Woody plant, growing as rounded clumps on the branches of trees such as poplar and apple. The yellow-green, curved leaves grow in pairs. Small flowers in spring are followed in winter by clusters of creamy-white fruits (6–10 mm). Cut stems are sold as Christmas decorations.

Main toxins
Lectins and viscotoxins

Risk
Few reported cases. Severe poisoning is unlikely to occur.

Symptoms
Ingestion may occasionally result in nausea, vomiting and diarrhoea.

HTA Category
n/a

Clump of mistletoe with unripe green berries, growing in a hawthorn tree

A branch with ripe berries (© RBG, Kew)

EUONYMUS
spindle tree

Family
Celastraceae

Description
Shrubs and small trees, evergreen or deciduous with attractive autumn colour, growing in woodland, and also widely cultivated. Some produce small flowers in spring and summer, followed by pink or pink-white, 4-lobed capsules (6–25 mm), that open to reveal up to 4 seeds (4–12 mm), each covered in an orange fleshy aril.

Main toxins
Cardiac glycosides and alkaloids

Risk
Very few reported cases. Ingestion is most likely to result in mild poisoning.

Symptoms
Ingestion can cause nausea, vomiting, diarrhoea, abdominal discomfort and choking. Ingestion of a large quantity may result in an irregular heart rhythm and decreased blood pressure.

HTA Category
C, Harmful if eaten

Euonymus europaeus, the pink fruit split to reveal orange arils around the seeds

Flowering *Euonymus europaeus*

TAXUS
yew

Family
Taxaceae

Description
Trees or shrubs, growing in the countryside and widely cultivated, often in church-yards. Leaves densely packed, small, straight, dark green. Small flowers produced in the spring. Female plants bear fleshy red or occasionally yellow 'fruits' (arils; 7–11 mm), each with a single, green seed (5–7 mm).

Main toxins
Taxane alkaloids

Risk
Many reported cases. Ingestion of small quantities may result in mild poisoning. Ingestion of large quantities may result in severe poisoning.

Symptoms
Ingestion of only the fleshy aril, which is non-toxic, is unlikely to result in symptoms. Ingestion of other parts, including chewed seeds, can cause vomiting, diarrhoea and abdominal pain after about 30–60 minutes. Dizziness, muscle weakness and lethargy may also occur. In severe cases, symptoms may include low blood pressure, breathing difficulties, convulsions and coma. Cardiac effects are also common. Acute anaphylaxis has occurred after ingestion of the leaves.

Contact from handling the wood has resulted in inflammation of the skin.

HTA Category
B, CAUTION toxic if eaten

Close-up of fruiting *Taxus baccata* 'Fastigiata'

Fleshy arils, cut open to show the seed
(© RBG, Kew)

CORIARIA
tutu

Family
Coriariaceae

Description
Small trees and large shrubs, occasionally grown in gardens. Arching branches bear simple leaves in pairs. Small flowers in spring are followed in summer by black, red or yellow fruits (5–12 mm).

Main toxins
Coriaria lactones

Risk
Very few reported cases. Ingestion may result in moderate poisoning.

Symptoms
Ingestion may result in vomiting and abdominal pain. Neurological symptoms including coma, visual difficulties, muscle tightness, convulsions, fever, agitation and disorientation may also occur. Blood and metabolic toxicity has been reported rarely.

HTA Category
B, CAUTION toxic if eaten

Coriaria myrtifolia, flowering

Fruiting branch of *Coriaria myrtifolia*

PHYTOLACCA
inkberry, pigeonberry, pokeroot, pokeweed

Family
Phytolaccaceae

Description
Usually robust garden plants, occasionally found in the wild. Tall, branching stems with large, simple leaves. Small, white or pink flowers in long spikes are followed by purple-black, shiny, juicy berries (6–10 mm), often divided into numerous segments.

Main toxins
Lectins and triterpene saponins

Risk
Very few reported cases. Accidental ingestion may result in moderate poisoning. Intentional ingestion may result in severe poisoning.

Symptoms
Ingestion may result in a burning sensation in the mouth and throat, and salivation. Nausea, vomiting, diarrhoea, abdominal pain, lethargy and weakness are common. Fluid loss may be severe, resulting in low blood pressure. When large amounts of raw leaves or fruit are eaten, more severe symptoms have been reported in North America, including breathing difficulties, convulsions and irregular heart rhythm.

Contact may result in skin irritation.

HTA Category
B, CAUTION toxic if eaten

Phytolacca clavigera fruit-head

Flowering Phytolacca americana

LANTANA
shrub verbena

Family
Verbenaceae

Description
Tender plants grown in the house or as summer bedding. Leaves simple, rough. Somewhat flat heads of bright-coloured flowers appear in the summer, the colour of the flowers changing towards the outside of the head. Green turning blue-black, metallic berries (2–7 mm), are sometimes produced.

Main toxins
Triterpenes

Risk
Very few reported cases. Ingestion may result in moderate poisoning.

Symptoms
Ingestion of parts other than ripe fruits may result in nausea, vomiting, diarrhoea, dilated pupils, lethargy, weakness, imbalance and visual intolerance to light, usually within 6 hours. In severe cases, difficult or laboured breathing, depressed tendon reflexes and coma have been reported.

Contact may result in skin irritation.

HTA Category
B, CAUTION toxic if eaten

Flowering and fruiting branch of *Lantana camara*

Lantana 'Lucky Lemon Cream' as a summer bedding plant

Fruit ripen to a metalic black (© RBG, Kew)

DAPHNE
mezereon, spurge laurel

Family
Thymelaeaceae

Description
Small to medium, deciduous and evergreen shrubs, sometimes found in woodland, also grown in gardens. Simple leaves. Small, tubular, green, pink, white or yellow, highly scented flowers are produced in the spring and summer, followed by black, red or yellow berries (6–14 mm).

Main toxins
Diterpenes (secotiglianes)

Risk
Very few reported cases. Ingestion may result in severe poisoning.

Symptoms
Ingestion of a small amount may cause a burning sensation in the mouth and throat, nausea, abdominal pain, vomiting and diarrhoea. Large ingestions may result in pallor, dilated pupils, severe vomiting and diarrhoea (which may be bloody), blistering of the stomach lining and convulsions. Some sources suggest that kidney damage may also occur.

Contact may result in skin irritation and possibly blistering.

HTA Category
B, CAUTION toxic if eaten; skin irritant

Daphne mezereum flowering in the spring

Daphne tangutica bearing a few red fruit

Daphne laureola with clusters of black fruit

PRUNUS LAUROCERASUS and P. LUSITANICA
cherry laurel, Portugal laurel

Synonyms
Cerasus, laurocerasus

Family
Rosaceae

Description
Evergreen shrubs and small trees, widely grown as hedges and ornamentals, also found in the wild. Leaves simple, leathery. Spikes of unpleasant-smelling, creamy-white flowers. Cherry-like fruits (10–15 mm; seeds 7–10 mm), ripen through red to black. Other species are grown for their edible fruit (almond, peach and plum).

Main toxins
Cyanogenic glycosides

Risk
Many reported cases. Accidental ingestion is unlikely to result in poisoning.

Symptoms
Ingestion occasionally causes mild gastrointestinal symptoms. Severe clinical effects are possible if the bitter-tasting seed kernels have been chewed.

HTA Category
C, Harmful if eaten (as *Prunus laurocerasus* and *P. lusitanica*)

Notes
Fresh leaves have the potential to form and release cyanide when damaged. Ensure plenty of ventilation when shredding fresh foliage, such as hedge trimmings, or when transporting them to a local composting facility.

Flowering *Prunus lusitanica* hedge, the leaves can be confused with those of *Laurus nobilis* (see page 31)

Prunus laurocerasus with hanging heads of ripe fruit

SAMBUCUS
danewort, dwarf elder, elder, elderberry

Family
Adoxaceae; sometimes referred to Caprifoliaceae

Description
Common small trees of hedgerows and waste ground, also grown in gardens. Leaves formed of several leaflets, sometimes finely toothed. Erect or drooping, flat or conical heads of small cream flowers in spring are followed in summer by small, black, purple-black, orange or red berries (4–8 mm).

Main toxins
Unknown and cyanogenic glycosides

Risk
Few reported cases. Ingestion may result in severe poisoning.

Symptoms
Ingestion of the raw fruits may result in nausea, vomiting and diarrhoea. Following ingestion of other parts of the plant, the onset of symptoms may be immediate or delayed, and include headache, a sensation of tightness in the throat and chest, palpitations and muscle weakness. In more severe cases there may be dizziness, difficulty in breathing, low blood pressure and collapse.

HTA Category
C, Harmful if eaten (except *Sambucus nigra*)

Sambucus nigra with hanging clusters of fruit

Sambucus nigra, flowering

Sambucus adnata has small, orange fruit

LIGUSTRUM
privet

Family
Oleaceae

Description
Usually evergreen shrubs or sometimes trees, growing in the countryside but also widely used as hedging. Simple leaves and the spikes of small, tubular, unpleasant-smelling, cream-white flowers are followed in autumn by small black berries (5–10 mm).

Main toxin
Unknown

Risk
Few reported cases. Ingestion of small quantities is likely to result in mild poisoning.

Symptoms
Ingestion can occasionally cause vomiting, abdominal pain and diarrhoea. There have been rare, questionable reports of severe poisoning in the older literature.

HTA Category
C, Harmful if eaten

Ligustrum vulgare, fruiting

Ligustrum ovalifolium, close-up of flowering hedge   *Ligustrum ovalifolium*, fruiting

RHAMNUS
alder buckthorn, black dogwood, buckthorn

Synonym
Frangula alnus

Family
Rhamnaceae

Description
Shrubs and small trees growing in thickets, hedges and fens, also sometimes cultivated. Stems sometimes thorny. Leaves simple, and can have attractive autumn colour. Clusters of small, yellowish-green flowers are followed in autumn by red, purple-black or black berries (6–13 mm).

Main toxins
Anthraquinones

Risk
Very few reported cases. Ingestion may result in moderate poisoning. Contact can result in mild poisoning.

Symptoms
Ingestion may result in nausea, vomiting, diarrhoea and abdominal pain. In severe cases, the diarrhoea may be violent and bloody with a risk of excess fluid loss and dehydration. The urine may be discoloured red. Use of *Rhamnus purshiana* as a herbal medication has resulted in liver poisoning. Death has been reported from ingestion of *R. cathartica* fruit, but this is very unusual.

Contact with thorns may cause mechanical damage, other plant parts may cause irritant contact dermatitis. Asthma and rhinitis have been reported from occupational exposure to *Rhamnus purshiana*.

HTA Category
C, Harmful if eaten

Fruiting *Rhamnus frangula*

Fruiting *Rhamnus tinctoria*

HEDERA
ivy

Family
Araliaceae

Description
Common, climbing and creeping plants, in the countryside, gardens and sold as houseplants. Leathery leaves, lobed or simple, green or variegated, remain on the plant all year. Can produce heads of cream flowers from late summer, followed by clusters of bitter tasting, black or occasionally yellow berries (4–10 mm).

Main toxins
Saponins and an allergen (falcarinol)

Risk
Many reported cases. Severe symptoms are unlikely to occur.

Symptoms
Ingestion commonly results in mild gastrointestinal upset only. More severe symptoms have been reported in the older literature after ingestion of leaves.

Contact can cause both irritant and, less commonly, allergic contact dermatitis. Several cases of severe skin irritation and blistering have been reported.

HTA Category
C, Harmful if eaten; may cause skin allergy

Fruiting plant; the fruit are similar to those of false castor oil plant, *Fatsia japonica* (page 30).

Flowering plant; young stems with lobed leaves are illustrated on page 16.

RHUS RADICANS
poison ivy

Synonym
Toxicodendron

Family
Anacardiaceae

Description
Only occasionally grown in Britain. These shrubs or woody climbers bear thin rather than leathery leaves with 3–7 leaflets. The leaves turn yellow, red or purple in the autumn before dropping. When damaged, the stems weep sap that turns black.

Main toxin
An allergen (urushiol)

Risk
Few reported cases. Ingestion may result in mild poisoning. Contact may result in severe poisoning.

Symptoms
Ingestion may result in nausea, vomiting and drowsiness.

Contact frequently results in severe dermatitis, with blistering, ulceration and staining or pigmentation of the skin. The initial sensitising reaction may take up to 3 weeks to manifest. Subsequent exposure may cause a reaction within 2–4 days that can persist for up to 3 weeks. Repeated exposure produces increasing severity of effects. Eye exposure may result in conjunctivitis and severe swelling of the eyelids. Inhalation of burning plant material may result in nausea, vomiting and swelling of the throat leading to difficulty in breathing.

HTA Category
A, CAUTION poisonous if eaten; skin contact commonly causes severe blistering dermatitis (only *Rhus radicans*, *R. succedanea* and *R. verniciflua*)

Notes
This does not include species of *Rhus* commonly known as sumachs, e.g. *R. typhina*.

Rhus radicans trained on a wall and showing autumn colour

FREMONTODENDRON
California beauty, flannel bush

Synonym
Fremontia

Family
Malvaceae (syn. Sterculiaceae)

Description
Large shrubs, usually trained against a wall or fence. Stems and lower leaf surfaces are covered with fine hairs that easily detach. Leaves lobed, leathery. Flowers large, yellow to orange, usually 5-petalled. Fruit a dry capsule.

Main toxin
None

Risk
Very few reported cases. Contact can result in moderate poisoning.

Symptoms
Ingestion has not been reported and is unlikely, but mild gastrointestinal upset is possible if this plant is eaten.

Contact with the small hairs can cause irritant contact dermatitis. Clouds of hairs released during pruning can irritate the eyes and respiratory tract. There is a possibility of allergic dermatitis in sensitive individuals.

HTA Category
C, Skin and eye irritant

Fremontodendron 'California Glory' trained on a trellis

The coating of brown hairs easily rubs off

FICUS CARICA
fig

Family
Moraceae

Description
Shrubs and small trees, often trained against a wall or fence, grown for their edible fruits and sometimes found in the wild. The large leaves are lobed and leathery. Green fruits (30–80 mm), are freely produced but often don't ripen to purple or brown.

Main toxins
Furanocoumarins

Risk
Very few reported cases. Ingestion or contact may result in mild poisoning.

Symptoms
Ingestion of fresh figs has occasionally resulted in a phototoxic reaction following exposure to sunlight.

Contact with bare skin in combination with exposure to sunlight may result in redness, swelling and burn-like lesions within 24 hours. Blisters may develop within 48 hours. Affected areas may be very itchy and painful. Effects subside within a few days but there may be subsequent brown pigmentation that can persist for several months, and affected areas can remain hypersensitive to ultraviolet light for several years. Eye contact has resulted in non-specific injury.

HTA Category
C, Skin irritant with sunlight

Leafy branches with unripe fruit

Unripe fruit cut open (© RBG, Kew)

AESCULUS
buckeye, conker tree, horse chestnut

Family
Sapindaceae (syn. Hippocastanaceae)

Description
Common, large trees and shrubs, widely grown, particularly in parks. Leaves large, hand-shaped. Upright heads of usually cream or pink flowers are followed in late summer and autumn by spherical, often spiny, capsules containing brown seeds (30–50 mm), known as conkers.

Main toxins
A coumarin glycoside (aesculin) and saponins (aescin)

Risk
Many reported cases. Ingestion of small quantities may result in mild poisoning.

Symptoms
Ingestion of the bitter seed may cause gastrointestinal upset with vomiting, abdominal pain and diarrhoea. Coma, hypertension and respiratory paralysis have been reported in a 4-year-old boy who repeatedly ingested seeds, but this is very unusual. Ingestion has also resulted in severe allergic reactions.

HTA Category
C, Harmful if eaten

Notes
The sweet chestnut (*Castanea sativa*), which produces edible seeds, is not included in this book.

Aesculus hippocastanum, fruiting branches

Flowering *Aesculus hippocastanum* tree

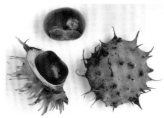

Conkers and spiny capsules (© RBG, Kew)

CANNABIS SATIVA

blow, dope, grass, hash, marijuana, pot, weed

Family
Cannabaceae

Description
Illegal drug plant and fibre crop, which occasionally grows on waste ground and in gardens from bird seed (3–6 mm). Hemp seed is also sold as fishing bait and as a dietary supplement. Plants can be stout but are often weedy. Leaves deeply divided.

Main toxins
Tetrahydrocannabinols

Risk
Many reports of ingestion (hashish resin) by children. Accidental ingestion can result in moderate poisoning.

Symptoms
Ingestion of the flowers or leaves may cause dizziness, unsteady gait, lack of coordination, drowsiness and tremor. Ingestion of hashish resin is associated with more severe effects, including increased heart rate and blood pressure, stupor, convulsions and coma.

HTA Category
n/a

Young leafy plants (© Robert Bevan-Jones)

Flowering *Cannabis sativa* (© E. Caballero)

Seeds from commercial bird seed (© RBG, Kew)

VITEX
chaste tree

Family
Lamiaceae (syn. Labiatae); sometimes referred to Verbenaceae

Description
Trees and shrubs grown in the garden for their aromatic foliage and fragrant flowers. Leaves divided like a hand into 3–7 leaflets. In late summer to autumn, small lilac to dark blue or white flowers are produced in large, branched heads.

Main toxin
Unknown

Risk
Very few reported cases. Ingestion or contact may result in mild poisoning.

Symptoms
Ingestion may result in mild gastrointestinal upset.

Contact may result in respiratory problems, nausea and dizziness, as well as in skin irritation, most frequently while cutting back and trimming the trees or shrubs.

HTA Category
C, Skin irritant

Vitex agnus-castus var. *latifolia*, flowering

Lower branches of *Vitex agnus-castus* f. *alba*

RICINUS COMMUNIS
castor bean plant, castor oil plant, palma Christi

Family
Euphorbiaceae

Description
Robust, tender garden plants. Leaves large with pointed lobes, grey-green or flushed red. Heads of small flowers are followed by green or red capsules, covered in soft spines. The large seeds (8–14 mm), are sold for home sowing. Also see false castor oil plant, *Fatsia japonica* (page 30).

Main toxin
A lectin (ricin)

Risk
Very few reported cases. Ingestion of chewed seeds may result in moderate to severe poisoning.

Symptoms
Ingestion of undamaged, intact seeds is unlikely to cause clinical effects because the hard seed coat prevents absorption. If seeds are chewed, symptoms, which may be delayed, include nausea, vomiting, abdominal pain, drowsiness, and liver and kidney problems.

Contact with the seeds can cause contact dermatitis and rarely severe allergic reactions.

HTA Category
B, CAUTION toxic if eaten

Flowering plants (© Robert Bevan-Jones)

Seed capsule and ripe seeds (© RBG, Kew)

Seeds imported into Britain on a necklace (© RBG, Kew)

EUPHORBIA (Outdoors)
spurge

Family
Euphorbiaceae

Description
Large group of small to large garden plants and weeds, all with white sap. Leaves usually linear or oval. From spring to early summer, insignificant flowers are surrounded by long-lasting, showy bracts, often in large heads. Seed capsules (2–20 mm), can be explosive when ripe.

Main toxins
Diterpene esters

Risk
Many reported cases. Ingestion or contact may result in mild to moderate poisoning.

Symptoms
Ingestion may result in nausea, salivation, vomiting and diarrhoea. There may be a burning sensation on the lips and tongue.

Contact may result in irritation, but reactions are not always immediate and vary in severity. Initially there is redness and itching, then swelling and possibly blistering. Blisters usually crust over in 1–3 days unless there is secondary infection. Recovery is usually complete within 4–7 days. Eye contact may result in severe irritation. Symptoms include pain, eye watering, swelling, decreased visual acuity, severe conjunctivitis and temporary blindness.

HTA Category
B, CAUTION skin and eye irritant; harmful if eaten (except *Euphorbia pulcherrima*, poinsettia)

Euphorbia lathyris can appear in unexpected places

Euphorbia characias flowers in the spring (© Mark Jackson)

Euphorbia griffithii 'Fireglow'

URTICA DIOICA
nettle, stinging nettle

Family
Urticaceae

Description
Wild plants of woodlands and hedges, and a weed of gardens and waste places. Leaves simple, toothed, throughout the stem. Leaves and stems are covered with stinging hairs. The small flowers are produced in late spring and summer, in drooping tassels. Young leaves are cooked and eaten.

Main toxins
Histamine, acetylcholine and serotonin

Risk
Very few reported cases. Accidental contact can result in mild effects.

Symptoms
Ingestion is associated with low toxicity and young plants are edible when cooked.

Contact can cause a pricking and tingling sensation on the affected area, the skin will redden and an itching rash develops.

HTA Category
n/a

Close-up of flowering plant

An expanse of flowering plants

RHEUM × HYBRIDUM
rhubarb

Synonyms
Rheum × cultorum, R. rhabarbarum

Family
Polygonaceae

Description
Robust plant grown in gardens and on allotments for its leaf stalks, which are cooked and eaten as a fruit. Leaves with large blades and long red-flushed stalks. Plants can be 'forced' by covering in the spring.

Main toxins
Anthraquinones and oxalates

Risk
Few reported cases. Ingestion may result in moderate poisoning.

Symptoms
Ingestion of leaves and uncooked leaf stems may cause effects within 2–12 hours. There may be a burning sensation in the mouth, throat and stomach, with vomiting (which may be bloody), abdominal pain, diarrhoea and drowsiness. Patients admitted to hospital are often dehydrated because of repeated vomiting. In severe cases, there may be muscle twitching, coma, convulsions, kidney and rarely liver damage. Recovery may take several days.

Contact may irritate the skin and can cause dermatitis.

HTA Category
n/a

Large leaves with pinkish-red stalks

Unforced rhubarb (© RBG, Kew)

ANTHRISCUS SYLVESTRIS

cow parsley, keck, Queen Anne's lace

Family
Apiaceae (syn. Umbelliferae)

Description
Common plant of hedges, the edges of woods and along roads. 'Ravenswing', a cultivar with purple-brown leaves, is sometimes sold. Heads of tiny white flowers are produced in spring above finely divided, bright green leaves.

Main toxins
Furanocoumarins

Risk
Very few reported cases. Contact may result in moderate poisoning.

Symptoms
Ingestion of a small amount is not expected to cause any adverse effects.

Contact with bare skin combined with exposure to sunlight may result in redness, swelling and burn-like lesions within 24 hours. Blisters may develop within 48 hours. Affected areas may be very itchy and painful. Effects subside within a few days but there may be subsequent brown pigmentation which can persist for several months, and affected areas can remain hypersensitive to ultraviolet light for several years.

HTA Category
n/a

A mass of flowering plants in the spring (© Mark Jackson)   Heads (umbels) of flowers and fruit

CONIUM MACULATUM
hemlock

Family
Apiaceae (syn. Umbelliferae)

Description
Very tall plants found in damp places and along roads. The purple-spotted stems and feathery, dark green leaves smell unpleasant when crushed. Flowers white, in flat heads, produced in late spring to early summer.

Main toxins
Piperidine alkaloids (coniine)

Risk
Very few reported cases. Ingestion may result in severe poisoning. Contact is likely to result in mild poisoning.

Symptoms
Ingestion is most likely to result in gastrointestinal upset. Rarely, there may be increases in heart rate, breathing rate and blood pressure, followed by sudden decreases in heart rate, breathing rate and blood pressure if large amounts are ingested. In severe cases, convulsions and muscular paralysis leading to respiratory failure can occur. Kidney poisoning has also been reported.

Contact may result in a burning sensation, numbness, dermatitis and possibly symptoms as for ingestion. Eye contact would be expected to result in irritation.

HTA Category
n/a

Tall flowering plant

Flower head seen from above

Purple-spotted stem

OENANTHE CROCATA
dead man's fingers, hemlock water dropwort

Family
Apiaceae (syn. Umbelliferae)

Description
Growing in wet soil, this wild plant has rounded heads of small, white flowers. Other species are occasionally cultivated, some of which are non-toxic (*Oenanthe javanica*, *O. pimpinelloides* and *O. sarmentosa*) whereas others are toxic (*O. aquatica* and *O. phellandrium*).

Main toxin
A polyacetylene alcohol (oenanthotoxin)

Risk
Very few reported cases. Ingestion may result in severe poisoning.

Symptoms
Ingestion may cause rapid onset of nausea, dizziness, dilated pupils, increased breathing and heart rate, drowsiness, severe convulsions and death.

HTA Category
B, CAUTION toxic if eaten (as *Oenanthe crocata*, *O. aquatica* and *O. phellandrium*)

Flowerhead (umbel)

Fresh, green, leafy stems of *Oenanthe crocata* just before flowering

The rootstock has been confused with parsnips (© RBG, Kew)

CICUTA VIROSA
cowbane, water cowbane, water hemlock

Family
Apiaceae (syn. Umbelliferae)

Description
Uncommon, wild plant of shallow, fresh water. The tap-roots have hollow chambers. Leaves large, with a long, hollow leaf stalk, the blade is formed of many leaflets. Flowers white, small, in flat-topped heads.

Main toxin
A polyacetylene alcohol (cicutoxin)

Risk
Very few reported cases. Ingestion may result in severe poisoning.

Symptoms
Ingestion may cause the rapid onset of a burning pain in the mouth, nausea, prolonged vomiting, dizziness, dilated pupils, increased breathing and heart rate, drowsiness and severe convulsions.

HTA Category
n/a

Plant growing at the edge of water

Flowerheads

Rootstock cut in half lengthways to show hollow chambers
(© RBG, Kew)

HERACLEUM

cow parsnip, giant hogweed, hogweed, keck

Family
Apiaceae (syn. Umbelliferae)

Description
Robust, wild and ornamental plants, often growing close to water. Stems hollow, ridged, sometimes spotted purple. Leaves large to very large, lobed, blunt or coarsely toothed. Large heads of white or pinkish flowers in late spring to early summer, the outer flowers larger.

Main toxins
Furanocoumarins

Risk
Very few reported cases. Contact may result in moderate effects.

Symptoms
Ingestion of a small amount is not expected to cause any adverse effects.

Contact with bare skin in combination with exposure to sunlight may result in redness, swelling and burn-like lesions within 24 hours. Blisters may develop within 48 hours. Affected areas may be very itchy and painful. Effects subside within a few days but there may be subsequent brown pigmentation that can persist for several months, and affected areas can remain hypersensitive to ultraviolet light for several years.

HTA Category
B, CAUTION severely toxic to skin with sunlight

Tall, flowering plants of *Heracleum sphondyllium*

Jagged-toothed leaves of *Heracleum mantegazzianum*

Heracleum mantegazzianum flowerhead

APIUM GRAVEOLENS
celeriac, celery, wild celery

Family
Apiaceae (syn. Umbelliferae)

Description
Wild plants in damp places, particularly by the sea; two varieties are cultivated as leaf (celery) and root (celeriac) vegetables. Stems grooved. Leaves coarsely divided, the lower ones with long leaf stalks. Heads of tiny white flowers are produced in summer.

Main toxins
Furanocoumarins, concentrations are higher in diseased leaves

Risk
Very few reported cases. Ingestion or contact may result in mild to moderate poisoning.

Symptoms
Ingestion of a large quantity in combination with exposure to ultraviolet light up to 8 hours later has resulted in poisoning. Symptoms may occur on any area of exposed skin and will follow those described for contact. Allergic reactions following ingestion include itching, swelling, abdominal cramps, dizziness, wheezing, difficulty in breathing and swollen tongue.

Contact with bare skin in combination with exposure to sunlight may result in redness, swelling and burn-like lesions within 24 hours. Blisters may develop within 48 hours. Affected areas may be very itchy and painful. Effects subside within a few days but there may be subsequent brown pigmentation that can persist for several months, and affected areas can remain hypersensitive to ultraviolet light for several years.

HTA Category
n/a

Celery plants on an allotment

Flowering celery with ridged stems (© Christine Leon)

PASTINACA SATIVA
parsnip, wild parsnip

Family
Apiaceae (syn. Umbelliferae)

Description
Wild plants of roadsides and waste places, also grown as a root crop. Tall, aromatic, finely hairy plants, dark or grey-green in colour. Leaves divided into leaflets. Heads of yellow flowers from late spring to summer.

Main toxins
Furanocoumarins

Risk
Very few reported cases. Contact may result in mild to moderate poisoning.

Symptoms
Ingestion is not expected to cause any adverse effects.

Contact with bare skin in combination with exposure to sunlight may result in redness, swelling and burn-like lesions within 24 hours. Blisters may develop within 48 hours. Affected areas may be very itchy and painful. Effects subside within a few days but there may be subsequent brown pigmentation that can persist for several months, and affected areas can remain hypersensitive to ultraviolet light for several years. Eye contact has resulted in non-specific injury.

HTA Category
n/a

Leaves and stems can be covered in short, soft hairs

Habit of naturalised parsnip cultivar (© Thomas H. F. Kidman)   Flowerheads with distinctive yellow flowers

HYPERICUM PERFORATUM
perforate St John's-wort, St John's-wort

Family
Clusiaceae (syn. Guttiferae)

Description
Wild flowers of hedgerows and rough grassland, sometimes grown in herb gardens. Leaves simple, with translucent, glandular dots. Branching heads of bright yellow flowers are produced in late spring to summer and are followed by dry seed capsules.

Main toxin
Hypericin

Risk
Very few reports of accidental exposure in humans. Ingestion of large quantities may result in mild poisoning.

Symptoms
Ingestion may result in mild gastrointestinal upset. Use of *Hypericum perforatum* extract may cause light-sensitive dermatitis.

Contact with skin is not thought to be a problem.

HTA Category
C, Harmful if eaten

Flowering plant

Head of flowers

CHELIDONIUM MAJUS
greater celandine

Family
Papaveraceae

Description
Quite common, somewhat large, wild plant, also occasionally cultivated. Sap orange. Leaves with rounded lobes of various sizes, the edges rounded-toothed. Yellow flowers with 4 or many petals, are produced in spring and are followed by long, slender capsules (30–75 mm).

Main toxins
Isoquinoline alkaloids

Risk
Very few reported cases. Ingestion of large quantities may result in moderate poisoning.

Symptoms
Ingestion is discouraged by the plant's unpleasant smell and taste. A burning sensation in the mouth and throat, vomiting, diarrhoea and abdominal discomfort are possible.

Eye contact with the orange milky sap can cause irritation and pain.

HTA Category
C, Harmful if eaten; skin and eye irritant

Flowering plant with distinctively lobed leaves

Close-up of a 4-petalled flower and several elongated seed pods

RUTA
garden rue, herb of grace, Jackman's blue, rue

Family
Rutaceae

Description
Small, evergreen shrubs and woody herbs, grown in gardens. Leaves grey- or blue-green, occasionally variegated with cream, with many rounded lobes. Flowers in loose branching heads, bright yellow, with 4 petals forming a cross. Fruit green, with 4 lobes.

Main toxins
Furanocoumarins, essential oils and alkaloids

Risk
Very few reported cases. Ingestion of a large amount may result in moderate poisoning. Contact may result in severe poisoning.

Symptoms
Ingestion of a large amount can cause severe gastric pain and vomiting.

Contact with bare skin in combination with exposure to sunlight may result in redness, swelling and burn-like lesions within 24 hours. Blisters may develop within 48 hours. Affected areas may be very itchy and painful. Effects subside within a few days but there may be subsequent brown pigmentation that can persist for several months, and affected areas can remain hypersensitive to ultraviolet light for several years.

HTA Category
B, CAUTION severely toxic to skin in sunlight

Flowering plant of *Ruta graveolens*

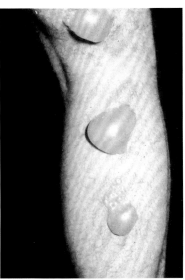

Contact with *Ruta* can result in a severe phototoxic reaction (© RBG, Kew)

DICTAMNUS ALBUS
burning bush, dittany

Family
Rutaceae

Description
Shrubby, aromatic plants occasionally grown in gardens. Leaves dark green, large, divided into 3–6 pairs of simple leaflets. Flowering in summer, flowers in long, erect heads, white or pink, with 5 petals, the stamens protruding and curved. Fruit with 5 lobes.

Main toxins
Furanocoumarins

Risk
Very few reported cases. Contact may result in mild to moderate effects.

Symptoms
Ingestion has not been reported, but mild gastrointestinal upset is possible if this plant is eaten.

Contact with bare skin in combination with exposure to sunlight may result in redness, swelling and burn-like lesions within 24 hours. Blisters may develop within 48 hours. Affected areas may be very itchy and painful. Effects subside within a few days but there may be subsequent brown pigmentation that can persist for several months, and affected areas can remain hypersensitive to ultraviolet light for several years. Eye contact has resulted in non-specific injury.

HTA Category
C, Skin irritant with sunlight

Flowering plant

A purple-flowered form

KALMIA
American laurel, calico bush

Family
Ericaceae

Description
Evergreen garden shrubs. Leaves simple, leathery. Bearing clusters of red, pink or white, bowl-shaped flowers formed of 5 fused petals, in late spring and early summer.

Main toxins
Grayanotoxins

Risk
Very few reported cases. Ingestion of small quantities may result in moderate poisoning.

Symptoms
Ingestion of flowers and contaminated honey has been reported in other countries. Initial effects are vomiting, a burning or numb sensation in the mouth, salivation and abdominal pain. Ingestion of a large quantity of plant material may result in slow heart rate, abnormally low blood pressure, irregular heart rhythm and neurological effects. Burning and itching of the skin has been reported after ingestion.

HTA Category
C, Harmful if eaten

Kalmia latifolia 'Pink Charm'

Kalmia latifolia with pale flowers and glossy leaves

Kalmia angustifolia has smaller flowers

RHODODENDRON
azalea

Synonym
Azalea

Family
Ericaceae

Description
Large group of evergreen and deciduous shrubs and small trees, widely grown in gardens and parks, and also found in the wild. Leaves variable, small to large, soft to leathery. Nectar-rich flowers produced in spring and summer in clusters of various colours, including red, pink, purple, orange, yellow and white.

Main toxins
Grayanotoxins, the amount varies between different species and cultivars

Risk
Few reported cases. Ingestion of small quantities is most likely to result in mild poisoning.

Symptoms
Ingestion can occasionally cause vomiting, diarrhoea and abdominal discomfort. More rarely, ingestion causes drowsiness, dizziness and decreased blood pressure and heart rate. Toxicity has been reported in Europe, North America and Asia following ingestion of honey produced by bees feeding on these plants.

Occupational contact can cause dermatitis.

HTA Category
n/a

Rhododendron ponticum, flowering shrub

Rhododendron × loderi has large, pale pink flowers

NERIUM OLEANDER
oleander, rose bay, rose laurel

Family
Apocynaceae

Description
Tender shrubs and small trees grown outside in sheltered spots or in a conservatory. Leaves dark or grey-green, elongated. Heads of pink, red, white or apricot flowers in summer, the flowers with 5 spreading petals or double. Fruit an elongated capsule.

Main toxins
Cardiac glycosides

Risk
Very few reported cases. Ingestion may result in severe poisoning.

Symptoms
Ingestion causes pain or numbness in the mouth, nausea, vomiting and diarrhoea (which may be blood stained). In severe cases, symptoms can include visual disturbances, and heart and kidney problems.

Contact can cause dermatitis, such as an irritant rash.

HTA Category
B, CAUTION toxic if eaten

Notes
The name oleander can also refer to *Thevetia* (page 147).

Close-up of flowerhead　　　　Flowering shrub

SPARTIUM JUNCEUM
Spanish broom

Family
Fabaceae (syn. Leguminosae)

Description
Large shrubs grown in gardens and parks, and sometimes found in the wild. Stems slender, dark green, with scattered, small leaves. Long heads of large yellow flowers in late spring to summer, followed by flattened, silky, bean-like pods (50–90 mm; seeds 4–5 mm) that split and twist open when dry.

Main toxins
Quinolizidine alkaloids

Risk
Very few reported cases. Ingestion is most likely to result in mild poisoning.

Symptoms
Ingestion may cause gastrointestinal upset. Abdominal pain, unsteady gait, numbness, tingling in the feet and a transient rash have been reported after ingestion of a large number of unripe fruit.

HTA Category
C, Harmful if eaten

Flowering branches

Flattened, bean-like pods

SOPHORA
kowhai, necklace pod

Family
Fabaceae (syn. Leguminosae)

Description
Shrubs and small trees sometimes grown in gardens, particularly in sheltered spots. Leaves with 2 rows of numerous leaflets. Bearing clusters of yellow, or occasionally blue and white, flowers in late spring. Fruit (30–120 mm; seeds 3–10 mm), are occasionally produced.

Main toxins
Quinolizidine alkaloids

Risk
Very few reported cases. Ingestion may result in mild poisoning.

Symptoms
Ingestion is most likely to result in gastrointestinal upset. Rarely, there may be increases in heart rate, breathing rate and blood pressure, followed by sudden decreases in heart rate, breathing rate and blood pressure if large amounts are ingested. Serious poisoning is extremely rare.

HTA Category
B, CAUTION toxic if eaten

Sophora microphylla 'Sun King', flowering in the spring   *Sophora microphylla* 'Sun King', single-seeded fruit

+LABURNOCYTISUS 'ADAMII'

Adam's laburnum, broom laburnum

Family
Fabaceae (syn. Leguminosae)

Description
Uncommon, small tree, which is a graft hybrid of *Laburnum anagyroides* (see page 119) and *Cytisus purpureus*, purple broom, a shrub with pink flowers. The hybrid flowers are yellow flushed pink and are infertile, but this plant can also produce the flowers and fruits of both parents.

Main toxins
Quinolizidine alkaloids

Risk
Very few reported cases. Ingestion may result in moderate poisoning.

Symptoms
Ingestion can cause nausea, vomiting, abdominal pain and diarrhoea. This can progress to headache, dizziness, confusion, dilated pupils, clammy skin, increased heart rate and temperature, difficulty in breathing and drowsiness.

HTA Category
B, CAUTION toxic if eaten

Yellowish-pink flowers of the graft hybrid

Small tree with flowers of both parents and the graft hybrid

LABURNUM
golden chain, golden rain

Synonym
Cytisus laburnum

Family
Fabaceae (syn. Leguminosae)

Description
Small trees widely grown in gardens, but also sometimes found in the wild. Each leaf has 3 leaflets. Hanging heads of bright yellow flowers are produced in late spring, followed by green pods (30–60 mm; see page 12), which dry to brown. Seeds (4–6 mm) ripen from green to brown and then black.

Main toxins
Quinolizidine alkaloids

Risk
Many reported cases. Ingestion of small quantities may result in moderate poisoning. Ingestion of large amounts may result in severe poisoning.

Symptoms
Ingestion may result in a burning sensation in the throat, nausea, vomiting, abdominal pain and occasionally diarrhoea. This can progress to headache, dizziness, confusion, dilated pupils, clammy skin, increased heart rate and temperature, difficulty in breathing and drowsiness. After ingesting large amounts, the more serious effects include convulsions, respiratory failure and coma.

HTA Category
B, CAUTION toxic if eaten

Laburnum anagyroides, flowering

Hanging bunches of bean-like pods (© Robert Bevan-Jones)

ROBINIA PSEUDOACACIA
acacia, black locust, false acacia

Family
Fabaceae (syn. Leguminosae)

Description
Large trees and shrubs, widely grown and occasionally found in the wild. Hanging clusters of cream or pink flowers in late spring and summer, are followed by bean-like pods (30–100 mm; seeds 4–10 mm). The flowers are cooked and eaten, and bees make "acacia" honey from them.

Main toxins
Lectins

Risk
Very few reported cases. Ingestion may result in moderate poisoning.

Symptoms
Ingestion will normally only result in nausea, vomiting, diarrhoea, abdominal pain and thirst within about 2 hours of ingestion. Occasionally, central nervous system effects, including dilated pupils, headache, dizziness, weakness, hyperexcitability or drowsiness, have been reported. Fever, bloody vomit, increased heart and breathing rates, liver and kidney dysfunction and coma are rare.

Contact with thorns may cause mechanical injury.

HTA Category
C, Harmful if eaten

Notes
A separate genus *Acacia* (mimosa) is not included in this book.

Hanging branch with heads of flowers

Ripening fruit-pods are flushed red

WISTERIA

Synonym
Wistaria

Family
Fabaceae (syn. Leguminosae)

Description
Vigorous woody climbers, often trained over house walls and pergolas. Leaves large with several leaflets. Hanging heads of mauve, white or pink, fragrant flowers are produced in summer. Large, bean-like, grey, hairy pods (50–200 mm; seeds 7–12 mm), are sometimes produced.

Main toxin
A lectin (wistarin)

Risk
Very few reported cases. Severe symptoms are unlikely to occur.

Symptoms
Ingestion of seeds may cause moderate gastrointestinal upset such as repeated vomiting. Other symptoms including headache, dizziness and confusion have been reported.

Contact with skin has caused redness and irritation.

HTA Category
C, Harmful if eaten

Wisteria floribunda, flowering

Wisteria floribunda 'Alba', flowering

Wisteria floribunda, fruiting

LUPINUS
lupin

Family
Fabaceae (syn. Leguminosae)

Description
Garden plants, also found on road verges, railway banks and in waste places. Leaves divided like a hand into many long leaflets. Tall spires of bright flowers in a variety of colours are followed by bean-like pods (40–70 mm) containing bitter seeds (3–5 mm). Seeds of the 'sweet lupin' are cooked and eaten in other countries.

Main toxins
Quinolizidine and other alkaloids

Risk
Few reported cases. Ingestion of a small amount may result in mild poisoning.

Symptoms
Ingestion occasionally causes gastrointestinal upset, including vomiting. More severe symptoms are possible following ingestion of large amounts.

HTA Category
C, Harmful if eaten

Lupinus 'Tom Tom the Piper's Son' with heads of flowers and developing fruit

Bean-like pods of Lupinus polyphyllus, with one split open to show seeds (© RBG, Kew)

LOBELIA
cardinal flower

Family
Campanulaceae

Description
Somewhat rare wild plants and widely grown herbaceous garden plants, some prefering damp conditions. Leaves simple. In summer, producing tall spikes of bright flowers with a prominent lip in blue, red, purple, pink and white. There is no evidence that the summer bedding lobelia (*Lobelia erinus*) is toxic.

Main toxins
Piperidine alkaloids

Risk
Very few reported cases. Accidental ingestion of plant material may result in mild poisoning.

Symptoms
Ingestion may result in irritation of the throat, nausea, vomiting and abdominal pain. Side effects from use as a herbal medicine include nausea, vomiting, coughing, tremor and dizziness. Overdose has caused more serious effects, rarely death.

Contact may result in irritant dermatitis.

HTA Category
C, Harmful if eaten (as *Lobelia* spp. except *L. erinus*)

Lobelia tupa, close-up of flowers

Lobelia 'Scarlet Fan', flowering plants

Lobelia siphilitica, close-up of flowers

DELPHINIUM
consolida, larkspur

Synonym
Consolida

Family
Ranunculaceae

Description
Garden perennials with a tap-root and annuals that are sown each year, occasionally found as a casual weed. Leaves more or less finely divided. In summer, bearing tall, loose or dense flower heads in white, pink, purple or blue. Fruit a dry, lobed capsule.

Main toxins
Diterpene alkaloids

Risk
Very few reported cases. Ingestion may result in severe poisoning.

Symptoms
Ingestion may result in nausea, vomiting, diarrhoea, abdominal pain, headache, numbness of mouth, lips and limbs, pins and needles and muscle weakness. Sweating, dizziness, confusion and increased frequency and/or depth of breathing may be seen. In severe cases, convulsions, physical restlessness and/or incoherence, irregular heart rhythm and coma may occur. Poisoning is far more common in animals than in humans.

Contact may result in dermatitis.

HTA Category
C, Harmful if eaten

Delphinium Belladonna Group

Larkspur 'Stock Flowered Mixed'

ACONITUM
aconite, helmet flower, monkshood, wolf's bane

Family
Ranunculaceae

Description
Occasionally growing on stream banks, also cultivated in gardens. Growing in spring from a tuberous root, the leaves are lobed and toothed. In summer, blue, purple, yellow, pink or white hooded flowers grow in dense or loose spikes. Fruit a dry, lobed capsule.

Main toxins
Diterpene alkaloids, particularly aconitine and mesaconitine

Risk
Very few reported cases. Ingestion of small quantities may result in severe poisoning.

Symptoms
Ingestion commonly results in nausea, vomiting, diarrhoea, abdominal pain, headache, numbness of mouth, lips and limbs, pins and needles and muscle weakness. There may also be sweating, dizziness, confusion and increased frequency and/or depth of breathing. In severe cases, convulsions, physical restlessness and/or incoherence, coma, low blood pressure and irregular heart rhythm occur.

Contact can result in irritation of exposed skin. Mild poisoning (nausea, headache and rapid heart beat) has also been reported.

HTA Category
B, CAUTION toxic if eaten; harmful through skin contact

Notes
Winter aconite is *Eranthis hyemalis*, a small, yellow-flowered plant not included in this book.

Violet-blue flowers of *Aconitum* 'Bressingham Spire'

Aconitum lycoctonum ssp. *vulparia* with yellow flowers

DIGITALIS
foxglove

Family
Plantaginaceae

Description
Common wild plant particularly on the edge of woods, also widely grown in gardens. Leaves simple, smooth or softly hairy (see page 22). Tall spires of tubular, purple, pink, white, yellow or orange flowers are produced in summer to early autumn. Fruit a dry capsule.

Main toxins
Cardiac glycosides

Risk
Many reported cases. Ingestion may result in severe poisoning.

Symptoms
Ingestion of flowers occasionally results in gastrointestinal effects. Ingestion of leaves can cause oral and abdominal pain, nausea, vomiting and diarrhoea. In severe cases, symptoms can include visual and perceptual disturbances and heart and kidney problems.

Contact with plant material can cause irritation.

HTA Category
B, CAUTION toxic if eaten

Typical purple flowers of *Digitalis purpurea*

Yellow-flowered *Digitalis grandiflora*

SYMPHYTUM
comfrey

Family
Boraginaceae

Description
Herbaceous plants of damp places, roadsides and hedgebanks, also grown in gardens or for 'green manure'. Leaves simple, softly or roughly hairy. Flowering in spring to summer on tall, leafy stems, flowers in curled heads, tubular, mauve, blue, pink, cream or white.

Main toxins
Pyrrolizidine alkaloids

Risk
Very few reported cases. Chronic ingestion may result in severe poisoning.

Symptoms
Ingestion: poisoning following acute (i.e. one-off) ingestion is unlikely, but is possible if a very large amount is eaten. Most reports of poisoning are due to chronic (long-term or regular) ingestion. Infants are at particular risk. Symptoms resulting from chronic exposure can take weeks, months or even years to develop. There may be nausea, vomiting and abdominal pain. In severe cases, there may be breathing difficulties, high blood pressure, liver and/or kidney failure and heart failure. Jaundice is seen occasionally.

Contact may result in mild irritation and dermatitis of the skin.

HTA Category
C, Harmful if eaten

Symphytum × *uplandicum*, Russian comfrey, flowering stem

White comfrey, *Symphytum orientale*

Flowers of common comfrey, *Symphytum officinale* (© Mark Jackson)

ECHIUM
viper's bugloss

Family
Boraginaceae

Description
Wild plants of grassy places, cliffs and sandy ground, also grown in gardens. Roughly hairy with rosettes of simple leaves. Spikes of small, funnel-shaped, blue, purple, pink, white or yellow flowers form in spring to early autumn.

Main toxins
Pyrrolizidine alkaloids

Risk
Very few reported cases. Severe symptoms are unlikely to occur.

Symptoms
Ingestion is generally associated with low toxicity. Chronic ingestion may cause severe liver damage.

Contact with short, bristly hairs on both the leaves and the stems can cause irritation, and severe inflammation and itching. A rash may develop on sensitized individuals in response to the abrasive hairs and possibly an allergen.

HTA Category
C, Skin irritant

Echium vulgare with vivid blue flowers

Densely flowering spikes of a cultivated form of *Echium vulgare*

AGROSTEMMA GITHAGO
corncockle

Synonym
Lychnis githago

Family
Caryophyllaceae

Description
Occasional, annual weeds of grain fields, now sold as wild flower seeds (2.5–4 mm) and grown in gardens and meadows. Stems tall, slender with simple leaves, all softly hairy. Flowers large, purple, pink or white, with long, leaf-like calyx lobes. Fruit an egg-shaped, dry capsule.

Main toxins
Saponins

Risk
Very few reported cases. Ingestion of small quantities may result in mild poisoning.

Symptoms
Ingestion may result in salivation, nausea, vomiting and diarrhoea. Eating contaminated bread has resulted in weakness, dizziness, headache and chills. Saponins may affect respiration and circulation, potentially leading to coma and convulsions.

HTA Category
C, Harmful if eaten

Flowering and fruiting plants

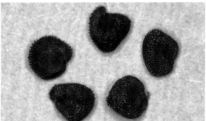

A single flower and young capsule

Seeds (© RBG, Kew)

PAPAVER SOMNIFERUM
opium poppy

Family
Papaveraceae

Description
Widely cultivated annual garden plant, also found on waste ground. Leaves large, coarsely and irregularly lobed. Short-lived flowers are produced in late spring and summer, each flower has 4 or many tissue-like petals. The ripe fruit (capsules; 25–90 mm) rattle when shaken, and the seeds (c. 0.5 mm) are used in baking.

Main toxins
Isoquinoline alkaloids

Risk
Few reported cases. Ingestion may result in severe poisoning.

Symptoms
Ingestion of the unripe fruit may result in drowsiness, constricted pupils, delirium, coma and shallow and slow breathing. In severe cases, there may be low blood pressure or low heart rate, convulsions and respiratory depression. Ingestion of other plant parts may result in gastrointestinal upset, twitching muscles and loss of bodily functions.

HTA Category
C, Harmful if eaten

Mature plants of *Papaver somniferum*

Close-up of flowers and capsule of a red cultivar

Capsules, cut to show numerous seeds (© RBG, Kew)

APOCYNUM
dogbane, Indian hemp

Family
Apocynaceae

Description
Rarely cultivated herbaceous perennials and shrubs with a milky sap. Leaves simple, in opposite pairs. Heads of small, whitish flowers are produced in the summer, followed by paired fruits. Each fruit is slender but up to 20 cm long, bursting open when ripe; seeds with silky hairs.

Main toxins
Cardiac glycosides

Risk
Very few reported cases. Ingestion is most likely to result in mild poisoning.

Symptoms
Ingestion can cause delayed-onset oral and abdominal pain, nausea, vomiting, diarrhoea and visual disturbance.

Contact with the milky sap can cause irritation.

HTA Category
C, Harmful if eaten

Apocynum cannabinum flowers (© Steven J. Baskauf)   *A. cannabinum*, flowering plant (© Steven J. Baskauf)

× CUPRESSOCYPARIS LEYLANDII
Leyland cypress

Family
Cupressaceae

Description
Vigorous, evergreen trees planted widely for hedging or as a screen. Leaves very small, scale-like, green, grey-green, yellow or variegated with cream. Flowers inconspicuous. Fruit a small cone.

Main toxin
Unknown allergen

Risk
Very few reported cases. Contact can result in mild effects.

Symptoms
Ingestion has not been reported, but mild gastrointestinal upset is possible if this plant is eaten.

Contact with skin, when planting or pruning hedges, can cause inflammation with a red, itchy rash. Sensitive individuals may develop a more severe reaction. Airborne contact dermatitis can be caused by smoke from burning branches.

HTA Category
C, May cause skin allergy

Notes
A hybrid between *Cupressus macrocarpa* and *Chamaecyparis nootkatensis*.

These trees have been left to grow rather than clipped into a hedge

The upper (left) and lower (right) surfaces of branch tips (© RBG, Kew)

CHRYSANTHEMUM
florists' chyrsanthemum

Synonym
Dendranthema

Family
Asteraceae (syn. Compositae)

Description
Popular and variable house and garden plants, also sold as cut flowers. Leaves usually dark green and variously lobed. Flowering in autumn, the 'flowers' formed from numerous florets, in yellow, orange, pink, white or green, singly or in branching heads.

Main toxins
Sesquiterpene lactones

Risk
Few reported cases. Contact may result in mild to moderate allergic reactions.

Symptoms
Ingestion has not been reported, but mild gastrointestinal upset is possible if this plant is eaten.

Occupational contact commonly results in dermatitis. Initially the fingers may be involved and later it may spread to the palms, forearms and face. A chronic dermatitis, particularly of the face, neck and backs of the hands, may develop. There may be a light-sensitive dermatitis-like distribution but the upper eyelids, behind the ears and the neck are also affected. In sensitive individuals, inhalation may result in rhinitis and eye exposure in conjunctivitis.

HTA Category
C, May cause skin allergy

Chrysanthemum 'Misty Lemon' and 'Misty Golden'   *Chrysanthemum* 'Mona Lisa Pink'

Indoor plants

Plants usually found growing inside

FICUS BENJAMINA
Benjamin tree, small-leaved rubber plant, weeping fig

Family
Moraceae

Description
Small trees commonly grown in houses and offices. Stems branching, slender and arching. Leaves somewhat small, simple, with an elongated tip, thinly leathery, dark or light green, grey-green or variegated with cream or white. Rarely flowers in Britain.

Main toxin
Unknown allergen

Risk
Few reported cases. Ingestion is most likely to result in mild poisoning. Contact can cause moderate effects.

Symptoms
Ingestion may result in mild gastrointestinal upset.

Contact may cause an allergic reaction with symptoms including a burning or itching sensation and a rash; the skin may appear swollen. Dust from the plant can cause conjunctivitis, nasal irritation, wheezing, coughing, chest tightness and shortness of breath.

HTA Category
C, May cause skin allergy

Ficus benjamina 'Danielle' with glossy, green leaves

Ficus benjamina 'Starlight' has variegated leaves

POLYSCIAS
dinner plate aralia

Family
Araliaceae

Description
Unusual shrubs and small trees grown as houseplants. Stems erect, branching infrequently, with leaves usually confined to the tips of the stems. Leaves simple or divided into leaflets, sometimes crinckled or with toothed edges. Rarely flowering or fruiting in Britain.

Main toxins
Saponins and a polyacetylene (falcarinone)

Risk
Very few reported cases. Ingestion and contact can result in mild poisoning.

Symptoms
Ingestion can cause immediate local irritation and inflammation of the mouth and throat. The irritation and bitter taste are likely to prevent significant ingestion.

Contact with plant material and its sap, particularly if prolonged, may cause local irritation, rash and generalised swelling.

HTA Category
C, Harmful if eaten; skin and eye irritant

Polyscias plants showing variation in leaf shape and colour, from left to right, *P. cumingiana*, *P.* 'Filicifolia' and *P. guilfoylei*

SCHEFFLERA
umbrella plant, umbrella tree

Synonym
Brassaia actinophylla

Family
Araliaceae

Description
Upright shrubs and small trees, grown as houseplants. Stems straight, branching infrequently. Leaves hand-shaped, formed of several leaflets, sometimes twisted, dark or light green, or variegated with cream or white. Large branching heads of small flowers are rarely produced.

Main toxins
An allergen (falcarinol) and possibly oxalates

Risk
Very few reported cases. Ingestion is most likely to result in mild poisoning.

Symptoms
Ingestion may result in a burning sensation in the mouth and mild gastrointestinal upset with vomiting and stomach pain.

Contact may cause dermatitis; these plants are known to be allergenic.

HTA Category
C, May cause skin allergy

Schefflera 'Compacta' with green leaves

Schefflera 'Trinette' is a variegated cultivar

PRIMULA OBCONICA
German primula, poison primula

Family
Primulaceae

Description
Popular houseplants, flowering in winter. With a basal rosette of long-stemmed, hairy, simple leaves. Bright flowers are produced in heads on long, hairy stems and may be red, pink, apricot, mauve or white, with a large leafy calyx at the back.

Main toxin
An allergen (primin)

Risk
Very few reported cases. Ingestion can result in mild poisoning. Contact can result in moderate poisoning.

Symptoms
Ingestion may result in gastrointestinal upset with vomiting, diarrhoea and abdominal pain. Chewing a leaf has caused an intense swelling of the mouth and throat.

Contact may result in allergic contact dermatitis. The fingers and face are most commonly affected, and there may be redness and swelling of the eyelids and around the mouth, neck, arms and hands. More rarely patches may appear on the ears, thighs, buttocks and ankles. The rash may cause a burning or itching sensation and there may be oedema (abnormal accumulation of fluid) and blistering. A distinctive pattern consisting of blotches and linear streaks is often seen, particularly on the arms and forearms.

HTA Category
B, CAUTION may cause skin allergy

A *Primula obconica* cultivar with orange-red flowers

Single plant with mauve flowers (© RBG, Kew)

ACALYPHA
cat's tail, chenille plant, copperleaf, Jacob's coat

Family
Euphorbiaceae

Description
Tender shrubs and annuals with a whitish sap (latex), grown as houseplants. Their attractive foliage is simple and may be twisted, it can be green or variously coloured with yellow, orange or brown. Some have red or cream, often hanging, catkin-like flower spikes.

Main toxins
Cyanogenic glycosides and diterpene esters

Risk
Very few reported cases. Ingestion and contact may result in mild poisoning.

Symptoms

Ingestion may cause irritation of the gastrointestinal system, but the reactions are immediate and therefore likely to be self-limiting.

Contact with the irritant sap could result in dermatitis. The reaction is immediate and is therefore likely to be self-limiting

HTA Category
C, Skin and eye irritant; harmful if eaten

Acalypha amentacea ssp. *wilkesiana* 'Macrophylla' with attractive foliage

Acalypha hispida has catkins of red flowers

BRUGMANSIA
angels' trumpets, tree datura

Family
Solanaceae

Description
Large shrubs or small trees that need winter protection but may be grown in the garden in summer. Leaves large, simple. Flowers very showy, hanging, trumpet-shaped or sometimes tubular, often fragrant in the evening, white, yellow, orange, pink or red. Fruit are elongated capsules that are rarely produced.

Main toxins
Tropane alkaloids

Risk
Very few reported cases. Accidental ingestion may result in moderate poisoning. Intentional ingestion can result in severe poisoning.

Symptoms
Ingestion can cause a dry mouth, blurred vision, warm and dry skin and dilated pupils. Other clinical effects include drowsiness, confusion, increased heart rate, reduced bowel sounds, difficulty in passing urine and disturbed vision or perception.

Contact can result in mild effects as above.

HTA Category
B, CAUTION toxic if eaten

Notes
See also the annual *Datura* (page 68).

Brugmansia arborea with white flowers

Pink-flowered *Brugmansia* × *insignis*

BRUNFELSIA
lady of the night, yesterday, today and tomorrow

Family
Solanaceae

Description
Uncommon tender shrubs and small trees grown as houseplants. Leaves simple. The fragrant flowers, which are flat with a short or long tube, start off purple and fade to nearly white over a few days; occasionally they are white becoming cream, or mauve darkening to purple. Fruit form rarely in Britain.

Main toxins
Pyrrole alkaloids

Risk
Very few reported cases. Ingestion may result in severe poisoning.

Symptoms
Ingestion may cause rapid onset of salivation, vomiting, diarrhoea, increased heart and respiration rates, disturbed vision or perception and seizures.

HTA Category
C, Harmful if eaten

Brunfelsia pauciflora 'Eximia', flowers become paler as they age

SOLANDRA
chalice vine, cup of gold

Family
Solanaceae

Description
Tender, vigorous, woody climbers that are sometimes grown in large conservatories. Leaves simple, leathery. Producing single, large, trumpet-shaped flowers that may be fragrant at night, yellow with purple veins, or white tinged with purple and ageing to yellow.

Main toxins
Tropane alkaloids

Risk
Very few reported cases. Ingestion may result in moderate poisoning.

Symptoms
Ingestion can cause dry mouth, blurred vision, dilated pupils, reduced bowel sounds, difficulty in passing urine and disturbed vision or perception. Increased heart rate and flushed face may also be present.

HTA Category
B, CAUTION toxic if eaten

Solandra maxima, young plant with a single flower (© Mark Jackson)

LAGENARIA
bottle gourd, calabash

Synonym
Cucurbita lagenaria

Family
Cucurbitaceae

Description
Tender annuals and perennials grown in a conservatory or greenhouse. Scrambling or climbing with tendrils. Flowers large, white. Fruits mature into hard-shelled, bottle-shaped or rounded gourds (up to 800 mm). The young fruits of *Lagenaria siceraria* (syn. *L. vulgaris*) can be cooked and eaten as a vegetable.

Main toxins
Saponins

Risk
Very few reported cases. Ingestion may result in moderate poisoning.

Symptoms
Ingestion of the fruits can cause severe vomiting and diarrhoea.

HTA Category
C, Harmful if eaten

Lagenaria siceraria flower (photo by P. Acevedo, courtesy of the Smithsonian Institution)

Lagenaria siceraria (© Forest & Kim Starr)

GLORIOSA SUPERBA
flame lily, glory lily

Family
Colchicaceae

Description
Scrambling and climbing conservatory plants, growing each year from an underground tuber. Leaves simple with the point elongated into a tendril. Showy flowers in summer and autumn, with six, wavy edged petals swept backwards, red, orange, yellow or a combination.

Main toxins
Colchicine alkaloids

Risk
Very few reported cases. Ingestion may result in severe poisoning. Contact may result in mild poisoning.

Symptoms
Ingestion may result in immediate irritation of the mouth. Other symptoms usually appear after a delay 2–12 hours. Common effects include nausea, vomiting, diarrhoea, abdominal pain, increased heart rate and chest pain. More severe effects such as low blood pressure, low heart rate, convulsions, irregular heart rhythm, and death have been reported. Blood, kidney and liver poisoning are also possible.

Contact may result in skin irritation.

HTA Category
B, CAUTION toxic if eaten

Gloriosa superba 'Rothschildiana' trained up a support

Gloriosa superba 'Lutea' is a yellow-flowered cultivar

GELSEMIUM
false yellow jasmine, evening trumpet flower, Carolina jessamine

Family
Gelsemiaceae

Description
Tender, evergreen, twining climber grown in conservatories. Leaves simple, glossy. In spring and summer, bearing clusters of fragrant pale or bright yellow flowers with an open, darker throat, or double.

Main toxins
Monoterpene indole alkaloids

Risk
Very few reported cases. Inappropriate use by adults can result in severe poisoning.

Symptoms
Ingestion can cause sweating, headache, weakness and difficulty in breathing. Severe poisonings have been reported following ingestion of herbal preparations containing the roots of *Gelsemium*.

HTA Category
C, Harmful if eaten

Gelsemium sempervirens (© Karl Gercens)

ALLAMANDA
golden trumpet

Synonym
Allemanda

Family
Apocynaceae

Description
Tender, robust, scrambling climbers with a milky sap. Leaves simple, leathery, in whorls of up to four. In summer, produces clusters or large, trumpet-shaped flowers with spreading lobes, usually yellow, but may be white, pink or peach.

Main toxin
An iridoid lactone (allamandin)

Risk
Very few reported cases. Severe symptoms are unlikely to occur.

Symptoms
Ingestion may cause nausea, abdominal pain and diarrhoea.

Contact with skin can cause an itchy, dry rash. Sap can cause an intense burning sensation in the eyes.

HTA Category
C, Harmful if eaten; skin and eye irritant

Allamanda cathartica 'Hendersonii' bears very large flowers   *Allamanda cathartica*, flowering plant

THEVETIA
yellow oleander

Synonyms
Cascabela thevetia, *Cerbera thevetia*

Family
Apocynaceae

Description
Conservatory shrubs and small trees. Leaves simple, narrow, leathery. Flowers from spring to autumn, fragrant, large, funnel-shaped, with yellow or apricot-yellow, overlapping petals, occasionally followed by a rounded-triangular fruit that ripens to black and contains 1–2 large seeds.

Main toxins
Cardiac glycosides

Risk
Very few reported cases. Ingestion may result in severe poisoning.

Symptoms
Ingestion can cause a burning sensation in the mouth, dryness in the throat, vomiting, diarrhoea, dilated pupils and irregular heart rhythm. Ingestion of the seeds can cause severe heart, kidney and liver toxicity.

Contact with the irritant sap can cause dermatitis, with a red, painful rash.

HTA Category
B, CAUTION toxic if eaten; skin irritant

Notes
Oleander is also used to refer to *Nerium* (page 115).

Flowering *Thevetia peruviana* in Hawaii (© Sonny Larsson)   Flower of *Thevetia peruviana* (© Sonny Larsson)

ACOKANTHERA
bushman's poison, poison bush, wintersweet

Synonym
Carissa acokanthera

Family
Apocynaceae

Description
Uncommon, tender, shrubs and small trees grown in the conservatory. Leaves simple, somewhat leathery. From winter to spring, produces clusters of small, white or pinkish, fragrant flowers, which are sometimes followed by plum-like fruits (20–25 mm) that ripen to black.

Main toxins
Cardiac glycosides

Risk
Very few reported cases. Ingestion may result in severe poisoning.

Symptoms
Ingestion can cause oral and abdominal pain, nausea, vomiting, diarrhoea, visual disturbances and severe changes in heart rhythm and blood pressure.

Contact may result in dermatitis.

HTA Category
C, Harmful if eaten

Notes
Wintersweet can also refer to other plants, e.g. *Chimonanthus praecox* (p. 27).

Acokanthera oblongifolia

Flowers of *A. oblongifolia*

Fruit of *A. oblongifolia*

TABERNAEMONTANA
crape or crepe jasmine, East Indian rosebay, pinwheel flower

Synonym
Ervatamia

Family
Apocynaceae

Description
Tender, shrubs and small trees, occasionally grown in conservatories. Leaves simple, dark green and glossy. Flowers are produced in small, branching heads, with 5 or more waxy petals, white with a yellow throat, fragrant in the evening.

Main toxins
Monoterpene indole alkaloids

Risk
Very few reported cases. Ingestion or contact may result in mild poisoning.

Symptoms
Ingestion of plant material can result in mild gastrointestinal upset. Ingestion of plant extracts can result in severe neurological and cardiac symptoms.
Contact can result in dermatitis.

HTA Category
C, Harmful if eaten

Tabernaemontana divaricata, with simple flowers

Tabernaemontana divaricata flowers can be double

ADENIUM
desert rose, impala lily, kudu lily, mock azalea, sabi star

Family
Apocynaceae

Description
Succulent houseplants with a white sap. The fleshy stems are thickened, particularly towards the base as the plant matures. Leaves are simple and glossy. Pink or white flowers, with an open throat and 5 spreading petals, appear in the summer.

Main toxins
Cardiac glycosides

Risk
Very few reported cases. Ingestion of small quantities may result in severe poisoning.

Symptoms
Ingestion of the latex or infusions of the plant can lead to severe poisoning, and may be fatal. After a delay of up to 6 hours, initial symptoms include oral and abdominal pain, nausea, vomiting, diarrhoea and visual disturbance. Irregular heart rhythm and low blood pressure may also occur.

HTA Category
C, Harmful if eaten

Adenium 'Sunshine' with fleshy stems and large flowers

Wild-collected Adenium obesum

The stems of mature Adenium obesum are swollen at the base

EUPHORBIA (Indoors)
crown of thorns, redbird cactus, slipper cactus, slipper spurge

Synonym
Pedilanthus

Family
Euphorbiaceae

Description
Succulent houseplants with a white sap. Leaves simple, green or variegated, restricted to the tips of the stems, scattered, absent or reduced in size. Flowers usually have showy bracts, which are often red, or they may be pointed like a bird's beak. See page 20 for poinsettia, *Euphorbia pulcherrima*.

Main toxins
Diterpene esters

Risk
Many reported cases. Ingestion or contact may result in mild poisoning.

Symptoms
Ingestion may result in oral irritation, protracted vomiting and diarrhoea, and abdominal pain.

Contact may result in mild skin irritation. Inflammation of the skin has been reported rarely. Eye contact may result in mild irritation.

HTA Category
C, skin and eye irritant; harmful if eaten (as *Pedilanthus*)

B, CAUTION skin and eye irritant; harmful if eaten (as *Euphorbia* except *E. pulcherrima*, poinsettia)

Euphorbia milii has thorny stems

Euphorbia tithymaloides (syn. *Pedilanthus tithymaloides*) with bird-like flowers

ALOE

Family
Xanthorrhoeaceae (syn. Aloaceae and Asphodelaceae)

Description
Usually tender plants, grown as houseplants particularly in the kitchen, in conservatories or outside in the summer. Leaves succulent, triangular, elongated, often toothed. Heads of flowers produced on long stems in the summer; flowers tubular, yellow, peach or orange.

Main toxins
Anthraquinones

Risk
Very few reported cases. Accidental ingestion may result in mild poisoning. Intentional ingestion may result in severe poisoning.

Symptoms
Ingestion may result in nausea, abdominal pain, vomiting and diarrhoea. Long-term use as a purgative (laxative) can cause loss of fluids and severe dehydration as well as damage to the colon, kidney malfunction and heart palpitations.

Contact may result in mild dermatitis.

HTA Category
C, Harmful if eaten

Aloe striatula can grow outside all year round

A typical houseplant, *Aloe vera*

OPUNTIA MICRODASYS
bunny ears

Family
Cactaceae

Description
Tender cactus grown as a houseplant. Stems flattened, oval to rounded oblong, branching, pale to mid-green, dotted with clusters of barbed bristles (glochids), which are usually yellow but may be white or reddish-brown. Flowers bright yellow, bowl-shaped.

Main toxin
None

Risk
Very few reported cases. Accidental contact can result in moderate effects.

Symptoms
Ingestion of significant quantities is unlikely due to the presence of barbed glochids (needle-shaped spines), which act as a deterrent.

Contact may result in skin discomfort, itching and a burning sensation, which may persist for up to a week. Small lesions can develop and usually clear within 2 weeks. Occasionally, the lesions may progress to inflammation, developing gradually up to 4 weeks post-exposure. Inflammation may persist for 2–8 months before resolving spontaneously. Eye exposure may result in inflammation, conjunctivitis, corneal erosion and inflammation of the cornea.

HTA Category
C, Skin irritant

Opuntia microdasys can grow into a large plant

Small plant of *Opuntia microdasys* f. *alba* with white glochids (© RBG, Kew)

ABRUS PRECATORIUS

coral bean, crab's eyes, jequerity or jequirity seeds, lucky beans, paternoster beans, prayer bean, rosary pea

Family
Fabaceae (syn. Leguminosae)

Description
Tropical climbers, very rarely grown in Britain, but the distinctive seeds are imported in jewellery, rosaries and musical instruments (e.g. inside maracas). The seeds (5–10 mm) are small, usually red or orange-red with a black spot, occasionally white.

Main toxin
A lectin (abrin)

Risk
Very few reported cases. Ingestion may result in severe poisoning.

Symptoms
Ingestion of seeds can cause severe poisoning, especially if they are chewed or damaged. There may be a latent period of 1–3 days before onset of symptoms. Ingestion may result in vomiting, watery or bloody diarrhoea, abdominal pain, drowsiness, weakness and increased blood pressure and heart rate. In severe cases, cardiac, liver and renal toxicity can occur.

Contact with skin is unlikely to cause symptoms. The seeds are irritant to the eye and may result in inflammation.

HTA Category
n/a

A pair of earrings imported from Peru, loose seeds and a necklace

AGLAONEMA
Chinese evergreens

Family
Araceae

Description
Houseplants with attractive leaves. Leaves simple, somewhat leathery, with an elongated tip, mid- or grey-green, patterned with silver or grey. Occasionally producing green 'flowers', which have a tubular spathe and cream spadix; followed by heads of red fruit (8–22 mm).

Main toxin
Calcium oxalate

Risk
Very few reported cases. Severe symptoms are unlikely to occur.

Symptoms
Ingestion of the leaves or sap may cause a burning sensation with redness, swelling and pain in the affected area. Mild gastrointestinal upset with nausea and vomiting can occur.

Contact with the sap is likely to be irritant. Eye exposure is likely to be irritant.

HTA Category
C, Harmful if eaten; skin and eye irritant

Typical houseplant, *Aglaonema* 'Maria Christina'

Flowering *Aglaonema commutatum* var. *maculatum*

Fruit (© RBG, Kew)

DIEFFENBACHIA
dumbcane, leopard lily

Family
Araceae

Description
Common houseplants with attractive leaves. Stems erect, cane-like. Leaves simple, pale or mid-green, often with white or cream markings. Occasionally flowering on short stems; 'flowers' with a tubular, green spathe and protruding cream spadix.

Main toxins
Calcium oxalate and dumbain

Risk
Few reported cases. Ingestion or contact may result in moderate poisoning.

Symptoms
Ingestion can cause immediate stinging, irritation and blistering of the lining of the mouth and throat, resulting in salivation, swelling and difficulty in swallowing. In rare cases, there may be airway obstruction, difficulty in breathing and impaired speech.

Contact with the sap may cause itching, blistering and dermatitis. Eye contact causes an immediate burning sensation, eye watering, twitching of the eyelids, sensitivity to light and swelling of the eyelids. Corneal injury varies from mild to severe. There may be visual disturbance for up to 3–8 weeks.

HTA Category
B, CAUTION toxic if eaten; skin and eye irritant

Mottled foliage of *Dieffenbachia maculata* 'Compacta'

Dieffenbachia seguine, flowering

SPATHIPHYLLUM
peace lily

Family
Araceae

Description
Common, flowering houseplants. Leaves with long stems, blades simple, thin-textured, dark green. 'Flowers' on long stems, spathe white, sometimes with green markings or entirely green, spadix knobbly, white or cream.

Main toxin
Calcium oxalate

Risk
Many reported cases. Ingestion or contact may result in mild poisoning.

Symptoms
Ingestion or chewing of plant material is expected to cause immediate stinging, irritation and blistering of mucous membranes in the mouth and throat, resulting in salivation and difficulty in swallowing.

Contact may result in urticaria. Inhalation may result in a runny nose, hoarseness or the complete loss of the voice, sore throat and asthma. Eye contact with the sap is likely to be irritant.

HTA Category
C, Harmful if eaten; skin and eye irritant

Spathiphyllum blandum, flowering plant

Close-up of *Spathiphyllum* cultivar, the individual flowers can be seen on the spadix

ANTHURIUM
flamingo flower, painter's palette, pigtail plant, tailflower

Family
Araceae

Description
Common, flowering houseplants, the 'flowers' also sold as cut flowers. Leaves with long stems, blades simple, leathery, dark green, sometimes with white veins. Brightly coloured, showy 'flowers' are produced freely; the spathe is bright or dull red, pink, or white, sometimes flushed green.

Main toxin
Calcium oxalate

Risk
Very few reported cases. Ingestion or contact are most likely to result in mild poisoning.

Symptoms
Ingestion of the plant or its sap can cause a burning sensation in the mouth and throat, other clinical effects are unlikely to occur.

Contact can cause irritation. A tingling sensation, pain, inflammation and persistent numbness of the local area have been reported. Eye exposure is likely to be irritant.

HTA Category
C, Harmful if eaten; skin and eye irritant

A pink *Anthurium* cultivar

Anthurium andraeanum with glossy red spathes

SYNGONIUM
arrowhead vine

Family
Araceae

Description
Creeping houseplants with attractive leaves. Stems slender, rooting from the joints. Leaves long-stemmed, arrow-shaped or with 3 leaflets, mid-green to dark green, with grey or silver-grey veins, or splashed with cream.

Main toxin
Calcium oxalate

Risk
Very few reported cases. Ingestion or contact is most likely to result in mild poisoning.

Symptoms
Ingestion of the plant or its sap can cause a burning sensation in the mouth and throat, other clinical effects are unlikely to occur.

Contact with skin and eyes can cause local irritation and inflammation.

HTA Category
C, Harmful if eaten; skin and eye irritant

A mature plant of *Syngonium auritum*

CALADIUM
angel's wings, elephant's ears

Family
Araceae

Description
Houseplants growing from a tuber and dying back in winter. Leaves with long stems, the blades simple, somewhat broadly arrow-shaped, attractively patterned on the veins or with a variety of splashes or spots in white, cream, red or pink.

Main toxin
Calcium oxalate

Risk
Very few reported cases. Ingestion or contact is most likely to result in mild poisoning.

Symptoms
Ingestion of the plant or its sap can cause a burning sensation in the mouth and throat, other clinical effects are unlikely to occur.

Contact with skin and eye can cause local irritation and inflammation.

HTA Category
C, Harmful if eaten; skin and eye irritant

Caladium 'Aaron'

Caladium 'Pink Beauty'

ALOCASIA
giant elephant's ear, giant taro

Synonyms
Colocasia indica, *C. macrorrhizos*

Family
Araceae

Description
Houseplants with large, long-stemmed, heart- to arrow-shaped leaves, which sometimes have veins of a contrasting colour. Giant taro (*Alocasia macrorrhizos*) is cultivated in the tropics for its rhizomes and shoots, which are edible if prepared adequately.

Main toxin
Calcium oxalate

Risk
Very few reported cases. Ingestion or contact is most likely to result in mild poisoning.

Symptoms
Ingestion of the plant or its sap can cause a burning sensation in the mouth and throat. Mild gastrointestinal upset with nausea and vomiting can occur if large quantities have been ingested.

Contact with skin and eye can cause local irritation and inflammation. Occupational allergic dermatitis has been reported.

HTA Category
C, Harmful if eaten; skin and eye irritant

Alocasia violacea likes damp conditions

Alocasia amazonica has leathery leaves with silver-grey veins

COLOCASIA ESCULENTA
cocoyam, dasheen, eddo, elephant's ears, taro

Family
Araceae

Description
Large-leaved houseplants or tender garden plants, requiring moist conditions. Leaves with long stems, the blades rounded arrow-shaped, fresh green or flushed very dark, reddish-brown. The tuber is grown as a vegetable in tropical countries but requires cooking.

Main toxin
Calcium oxalate

Risk
Very few reported cases. Ingestion or contact is most likely to result in mild poisoning.

Symptoms
Ingestion of the uncooked plant or its sap can cause local irritation and inflammation.

Contact with skin and eyes may cause irritation.

HTA Category
C, Harmful if eaten; skin and eye irritant

Colocasia esculenta likes to grow in moist conditions   *Colocasia gigantea*

XANTHOSOMA
blue taro, Indian kale, yautia

Family
Araceae

Description
Moisture-loving, house and conservatory plants. Leaves very large, stems long, blades broadly rounded, arrow-shaped, green or flushed dark purple. Rarely flowers or fruits in Britain. The leaves and tubers of some cultivars are edible if prepared adequately.

Main toxin
Calcium oxalate

Risk
Very few reported cases. Ingestion or contact may result in mild poisoning.

Symptoms
Ingestion is likely to cause immediate irritation in the mouth and a burning sensation in the throat. Mild gastrointestinal upset with nausea and vomiting can occur.

Contact with the sap may result in mild to moderate irritation of the skin and eyes.

HTA Category
C, Harmful if eaten; skin and eye irritant

The leaf stalks of *Xanthosoma violaceum* are dark purple

Xanthosoma sagitifolium, flowering

Xanthosoma sagitifolium, fruiting

RHAPHIDOPHORA
shingle plant

Synonym
Raphidophora

Family
Araceae

Description
Climbing and creeping house and conservatory plants, rooting at the joints of the cane-like stems. Leaves simple or divided, glossy, dark green. Rarely flowering in Britain.

Main toxin
Calcium oxalate

Risk
Very few reported cases. Ingestion or contact are most likely to result in mild poisoning.

Symptoms
Ingestion of the plant or its sap can cause a burning sensation in the mouth and throat, other clinical effects are unlikely to occur.

Contact with the plant or its sap can cause local irritation and inflammation.

HTA Category
C, Harmful if eaten; skin and eye irritant

Notes
Rhaphidophora aurea is a synonym of *Epipremnum aureum* (page 165)

Rhaphidophora petrieana, a narrow-leaved species

Rhaphidophora latevaginata, flowering

EPIPREMNUM
Devil's ivy, golden pothos, money plant

Synonyms
Pothos aureus, Rhaphidophora aurea, Scindapsus aureus

Family
Araceae

Description
Climbing or trailing houseplants, rooting along the stem. Leaves simple, heart-shaped, smooth or satin-textured, bright or dark green, often marbled or splashed with yellow or cream, or entirely yellow.

Main toxin
Calcium oxalate

Risk
Few reported cases. Ingestion or contact may result in mild poisoning.

Symptoms
Ingestion may result in immediate local irritation in the mouth and a burning sensation in the throat, which may become swollen. Mild gastrointestinal upset with nausea and vomiting can occur.

Contact may result in mild skin or eye irritation and mild allergic dermatitis.

HTA Category
C, Harmful if eaten; skin and eye irritant

Epipremnum aureum trained up a post

Hanging houseplant, *Epipremnum pictum* 'Argyraeum'

PHILODENDRON
blushing philodendron, heart leaf, sweetheart plant

Family
Araceae

Description
Houseplants that root along the stems and climb with support. Leaves simple, heart-shaped or more elongated, sometimes large and deeply lobed, glossy, leathery, mid- to dark green, sometimes flushed red on leaf stalks and stem.

Main toxins
Calcium oxalate and an allergen

Risk
Very few reported cases. Ingestion or contact is most likely to result in mild poisoning.

Symptoms
Ingestion of the plant or its sap can cause a burning sensation in the mouth and throat, other clinical effects are unlikely to occur.

Contact with skin and eyes can cause local irritation and inflammation. Allergic reactions have been reported following prolonged or repeated contact.

HTA Category
C, Harmful if eaten; skin and eye irritant

Typical houseplant, *Philodendron erubescens* (© RBG, Kew) *Philodendron bipinnatifidum* has deeply divided leaves

MONSTERA DELICIOSA
cheese plant, Swiss cheese plant

Family
Araceae

Description
Robust, climbing houseplants that root along the stems. Leaves with long stems, blades large, heart-shaped, entire on young plants, those on older plants with slits and holes. Rarely flowering in Britain. The fruit is edible when ripe and tastes of pineapple.

Main toxin
Calcium oxalate

Risk
Few reported cases. Ingestion or contact can result in mild poisoning.

Symptoms
Ingestion may cause irritation in the mouth but severe clinical effects are unlikely.

Contact with the plant or its sap can cause local irritation and inflammation.

HTA Category
C, Harmful if eaten; skin and eye irritant

Typical houseplant

A variegated form of *Monstera deliciosa* growing with other climbers

Sources of further information

Poisonous plants

Dauncey, E. A. (ed) (2000). *Poisonous Plants and Fungi in Britain and Ireland.* Interactive Identification Systems on CD-ROM. Royal Botanic Gardens, Kew and Guy's & St Thomas' NHS Foundation Trust, London.

Cooper, M. R., Johnson, A. W. & Dauncey, E. A. (2003). *Poisonous Plants and Fungi: an Illustrated Guide.* TSO, London.

Cooper, M. R. & Johnson, A. W. (1998). *Poisonous Plants and Fungi in Britain: Animal and Human Poisoning.* TSO, London.

Bevan-Jones, R. (2009). *Poisonous Plants: a Cultural and Social History.* Windgather Press, Oxford.

Royal Horticultural Society (1998). *Conservation and Environment Guidelines: Potentially Harmful Garden Plants.* [Leaflet.] Science Department, RHS, Wisley.

Crosby, D. G. (2004). *The Poisoned Weed: Plants Toxic to Skin.* Oxford University Press, New York.

Lovell, C. R. (1993). *Plants and the Skin.* Blackwell Scientific Publications, Oxford.

Other subjects

Rose, F. & O'Reilly, C. (2006). *The Wild Flower Key: How to Identify Wild Flowers, Trees and Shrubs in Britain and Ireland.* Frederick Warne, London.

Stace, C. (1997). *New Flora of the British Isles*, 2nd Ed. Cambridge University Press, Cambridge.

McVicar, J. (1997). *Good Enough to Eat: Growing and Cooking Edible Flowers.* Kyle Cathie Limited, London.

Ogren, T. L. (2000). *Allergy-Free Gardening: The Revolutionary Guide to Healthy Landscaping.* Ten Speed Press, Berkeley.

Huntington, L. (1998). *Creating a Low-Allergen Garden.* Mitchell Beazley, London.

Brostoff, J. & Gamlin, L. (2008). *The Complete Guide to Food Allergy and Intolerance*, 4th Ed. Quality Health Books, London.

Brostoff, J. & Gamlin, L. (1997). *Hayfever: The Complete Guide*, New Ed. Bloomsbury Publishing, London.

National Pollen and Aerobiology Research Unit: www.pollenuk.co.uk

General garden safety advice is available from the Royal Horticultural Society at: http://www.rhs.org.uk/Children/For-families/Safety-tips

Enquiries

Requests for information on poisonous plants may be sent (in writing only) to the Centre for Economic Botany, Royal Botanic Gardens, Kew, Richmond, Surrey TW9 3AB. Email: ceb-enq@kew.org; Fax: 020 8332 3717.

Our website is at http://www.kew.org/poisonous-plants

Urgent enquiries only can be made by telephone to 020 8332 5000 (office hours, Monday–Friday).

Acknowledgements

Photographs have been taken by the author except for those listed below. We would like to thank the following people for kindly allowing us to use their images: P. Acevedo (*Lagenaria siceraria* p. 143), Steven J. Baskauf (*Apocynum cannabinum* p. 131), Robert Bevan-Jones (*Cannabis sativa* p. 96, *Laburnum* p. 119, *Ricinus communis* p. 98, *Veratrum album* p. 52), E. Caballero (*Cannabis sativa* p. 96), Karl Gercens [www.KarlGercens.com] (*Gelsemium sempervirens* p. 145), Mark Jackson (*Anthriscus sylvestris* p. 102, *Euphorbia characias* p. 99, *Gaultheria mucronata* p. 78, *Iris foetidissima* p. 42, *Lysichiton americanus* p. 40, *L. camtschatensis* p. 40, *Narcissus* p. 58, *Passiflora caerulea* p. 49, *Solandra maxima* p. 142, *Symphytum officinale* p. 127), Thomas H. F. Kidman (*Pastinaca sativa* p. 108), Christine Leon (*Apium graveolens* p. 107), Dr Sonny Larsson (*Calla palustris* p. 39, *Thevetia peruviana* p. 147), Forest & Kim Starr [www.hear.org/starr] (*Lagenaria siceraria* p. 143) and Waterside Nursery [www.watersidenursery.co.uk] (*Calla palustris* p. 39).

Squires Garden Centre, Twickenham, allowed me to photograph plants; my thanks to Louise Jackson. Many Exhibitors at the Hampton Court Palace Flower Show, organised by the Royal Horticultural Society, let me photograph their plants, and I would particularly like to thank Raine Clarke-Wills (www.rainegardendesign.co.uk) and Fiona Godman-Dorington for free access to their show garden, 'The Dark Side of Beauty', which is illustrated on page 7, 23 and 34.

The toxicity information in this book is based largely on accounts written for the CD-ROM 'Poisonous Plants and Fungi in Britain and Ireland', published by Kew in 2000, and an unpublished Horticultural Trades Association Report by: Marion Cooper, Anthony Johnson and Specialists in Poisons Information of the Medical Toxicology Information Services, in particular Nicola Bates, but also Peter Barber, Jennifer Butler, Mark Colbridge, Grainne Cullen, Digby Green, Robie Kamanyire, Frances Northall, Marie Pickford, Elizabeth Schofield and Nicola Scott. Nick Edwards and Alison Dines advised and encouraged. The introduction was greatly improved by Heather Wiseman.

At the Royal Botanic Gardens, Kew: Jill Turner researched the non-toxic plant list and has provided assistance in numerous ways; Christine Leon was always supportive; Dr Sonny Larsson advised on plant names and toxins; Dr Pamela Aihie, Simon Baddeley, Geoffrey Butcher and Anne Edmonds provided reference support; the skilled horticultural staff not only grew many of the plants photographed for this book, but were also very helpful in locating specimens.

I am grateful to my family for their constant encouragement, patience and support during the preparation of this book.

Key to berry and flower colour

	Berry colour								Flower colour								
	Red	Orange	Yellow	Pink	Brown	Blue	Black	White	Green	Red	Orange	Yellow	Cream	Pink	Purple	Lavender	Blue	Black	White	Green
Abrus precatorius (p. 154)																			✓	
Acalypha (p. 139)																			✓	
Acokanthera (p. 148)							■												✓	
Aconitum (p. 125)															■				✓	
Actaea section Actaea (p. 41)							■	✓											✓	
Adenium (p. 150)														■					✓	
Aesculus (p. 95)																			✓	
Aglaonema (p. 155)	■																		✓	
Agrostemma githago (p. 129)														■					✓	
Allamanda (p. 146)												■							✓	
Alocasia (p. 161)		■																	✓	■
Aloe (p. 152)																			✓	
Alstroemeria (p. 64)																			✓	
Amaryllis belladonna (p. 61)																			✓	
Anthriscus sylvestris (p. 102)																			✓	
Anthurium (p. 158)	■																		✓	
Apium graveolens (p. 107)																			✓	
Apocynum (p. 131)														■					✓	
Arisaema (p. 36)	■														■				✓	
Arum (p. 35)	■									■								■	✓	
Asparagus (p. 45)	■																		✓	
Atropa belladonna (p. 71)							■												✓	
Aucuba japonica (p. 29)	■														■				✓	
Berberis (p. 29)	■						■												✓	
Brugmansia (p. 140)																			✓	
Brunfelsia (p. 141)															■				✓	
Bryonia dioica (p. 46)	■																		✓	
Caladium (p. 160)								✓											✓	
Calla palustris (p. 39)	■																		✓	
Cannabis sativa (p. 96)																				
Capsicum annuum (p. 77)	■	■					■								■				✓	
Chelidonium majus (p. 110)																				
Chrysanthemum (p. 133)												■							✓	
Cicuta virosa (p. 105)																			✓	
Colchicum (p. 62)															■				✓	
Colocasia esculenta (p. 162)									■										✓	
Conium maculatum (p. 103)																			✓	
Convallaria (p. 43)	■																		✓	
Coriaria (p. 83)							■													
Cotoneaster (p. 29)	■																		✓	
Crataegus (p. 30)	■																		✓	
X *Cupressocyparis leylandii* (p. 132)																				
Daphne (p. 86)	■							✓							■				✓	
Datura (p. 68)																			✓	
Delphinium (p. 124)															■		■		✓	
Dictamnus albus (p. 112)														■					✓	
Dieffenbachia (p. 156)																			✓	■
Digitalis (p. 126)														■						
Dioscorea communis (p. 47)	■																			
Dracunculus (p. 37)															■					
Echium (p. 128)														■			■		✓	
Epipremnum (p. 165)																				
Euonymus (p. 81)		■		■				✓											✓	
Euphorbia (pp. 99, 151)																			✓	
Fatsia japonica (p. 30)							■												✓	
Ficus benjamina (p. 135)	■																			
Ficus carica (p. 94)				■																
Fremontodendron (p. 93)											■									
Fuchsia (p. 30)				■										■					✓	
Gaultheria mucronata (p. 78)								✓											✓	
Gelsemium (p. 145)																				
Gloriosa superba (p. 144)												■							✓	
Hedera (p. 91)							■												✓	
Helleborus (p. 51)																			✓	
Heracleum (p. 106)																			✓	
Hippeastrum (p. 60)										■									✓	
Hyacinthoides (p. 56)																■	■		✓	
Hyacinthus (p. 57)														■					✓	
Hyoscyamus (p. 67)																			✓	
Hypericum (p. 30)							■					■								
Hypericum perforatum (p. 109)																				
Ilex (p. 30)	■							✓											✓	
Ipomoea (p. 48)															■				✓	
Iris (p. 42)								✓				■					■		✓	

Plant	Berry colour								Flower colour								
	Orange	Yellow	Pink	Red	Blue	Black	White	Green	Orange	Yellow	Cream	Pink	Lavender	Blue	Black	White	Green
Kalmia (p. 113)																✓	
+*Laburnocytisus* 'Adamii' (p.118)																	
Laburnum (p. 119)																	
Lagenaria (p. 143)																✓	
Lantana (p. 85)																✓	
Laurus nobilis (p. 31)						■										✓	
Ligustrum (p. 89)						■										✓	
Lobelia (p. 123)																✓	
Lonicera (p. 31)	■			■		■										✓	
Lupinus (p. 122)																✓	
Lycium (p. 31)				■													
Lysichiton (p. 40)																✓	
Mahonia (p. 31)					■												
Mandragora (p. 69)																✓	
Mirabilis (p. 65)																✓	
Monstera deliciosa (p. 167)							✓									✓	
Moraea section *Homeria* (p. 63)																✓	
Narcissus (p. 58)																✓	
Nerium oleander (p. 115)																✓	
Nicotiana (p. 66)																✓	
Oenanthe crocata (p. 104)																	
Opuntia microdasys (p. 153)	■			■													
Ornithogalum (p. 54)																✓	
Papaver somniferum (p. 130)															■		
Passiflora (p. 49)																✓	
Pastinaca sativa (p. 108)																	
Philodendron (p. 166)																✓	
Physalis alkekengi (p. 31)																✓	
Phytolacca (p. 84)						■										✓	
Podophyllum (p. 50)																✓	
Polygonatum (p. 44)						■										✓	
Polyscias (p. 136)						■										✓	
Primula obconica (p. 138)																✓	
Prunus (p. 87)																✓	
Pyracantha (p. 32)																✓	
Rhamnus (p. 90)																✓	
Rhaphidophora (p. 164)																	
Rheum x hybridum (p. 101)																	
Rhododendron (p. 114)																✓	
Rhus radicans (p. 92)																✓	
Ricinus communis (p. 98)																	
Robinia pseudoacacia (p. 120)																✓	
Ruta (p. 111)																	
Sambucus (p. 88)						■	✓									✓	
Schefflera (p. 137)																✓	
Scilla (p. 55)																✓	
Scopolia (p. 70)																	
Solandra (p. 142)																✓	
Solanum dulcamara (p. 75)																	
Solanum nigrum (p. 72)						■										✓	
Solanum pseudocapsicum (p. 76)																✓	
Solanum tuberosum (p. 73)																✓	
Solanum (other species) (p. 74)	■			■		■	✓									✓	
Sophora (p. 117)																✓	
Sorbus (p. 32)	■			■			✓									✓	
Spartium junceum (p. 116)																	
Spathiphyllum (p. 157)																✓	
Symphoricarpos (p. 79)			■				✓									✓	
Symphytum (p. 127)																✓	
Syngonium (p. 159)																✓	
Tabernaemontana (p. 149)																✓	
Taxus (p. 82)																	
Thevetia (p. 147)																✓	
Tulipa (p. 59)																✓	
Urtica dioica (p. 100)																	
Vaccinium corymbosum (p. 32)					■											✓	
Veratrum (p. 52)															■	✓	
Viscum album (p. 80)							✓										
Vitex (p. 97)																✓	
Wisteria (p. 121)																✓	
Xanthosoma (p. 163)																✓	
Zantedeschia (p. 38)																✓	
Zigadenus (p. 53)																✓	

Index

Name – name of a 'Plant Profile' in this book
Name – other scientific (Latin) name
name – common name
'Name' – cultivar
Page numbers in **bold** indicate images and main entries

Abrus precatorius 33, **154**
Acacia 120
acacia **120**
acacia, false **120**
Acalypha **139**
  amentacea ssp. *wilkesiana*
    'Macrophylla' **139**
  hispida **139**
Achillea 17
Acokanthera **148**
  olblongifolia **148**
aconite **125**
aconite, winter 125
Aconitum 20, 33, **125**
  lycoctonum ssp. *vulparia* **125**
  'Bressingham Spire' **125**
Actaea *pachypoda* **41**
  racemosa 41
  rubra ssp. *arguta* **41**
    f. *neglecta* **41**
  section *Actaea* **41**
  spicata 41
Adam's laburnum 33, **118**
Adenium **150**
  obesum **150**
  'Sunshine' **150**
Adoxaceae 88
Aesculus **95**
  hippocastanum **95**
Agastache 26
Agave 18
  parryi **19**
Ageratum 25
Aglaonema **155**
  commutatum var. *maculatum* **155**
  'Maria Christina' **155**
Agrostemma githago 13, **129**
alder buckthorn **90**
Allamanda **146**
  cathartica **146**
    'Hendersonii' **146**
Allemanda 146
Allium 17, 25
almond 87
Aloaceae 152
Alocasia **161**
  amazonica **161**
  macrorrhizos 161
  violacea **161**
Aloe 18, **152**
  striatula **152**
  vera **152**
Alstroemeria 16, 17, **64**
  aurea **64**
  'Tessa' **64**
Alstroemeriaceae 64
Amaranthus caudatus 25
Amaryllidaceae 58, 60, 61
Amaryllis belladonna **61**

amaryllis **60**, 61
Amberboa moschata 26
American laurel **113**
Anacardiaceae 92
Anemone patens 17
angel's wings **160**
angels' trumpets 33, **140**
Anigozanthos 17
Anthriscus sylvestris **10**, 14, **102**, 169
Anthurium 20, **158**
  andraeanum **158**
Antirrhinum 25
Apiaceae 12, 14, 102, 103, 104, 105, 106, 107, 108
Apium graveolens 14, **107**, 169
Apocynaceae 115, 131, 146, 147, 148, 149, 150
Apocynum **131**
  cannabinum **131**, 169
apple 11
apple, kangaroo **74**
  May **50**
Aquilegia 27
Arabis 27
Araceae 14, 20, 35, 36, 37, 38, 39, 40, 155, 156, 157, 158, 159, 160, 161, 162, 163, 164, 165, 166, 167
Arachnioides adiantiformis 17
aralia, dinner plate **136**
Araliaceae 91, 136, 137
Arisaema **36**
  ciliatum var. *liubaense* **36**
  tortuosum **36**
arrowhead vine **159**
Arum 33, **35**
  italicum **35**
  maculatum **35**
arum family 20
  lily **38**
arum, bog **39**
  dragon **37**
  water **39**
Asclepias curassavica 17
Asparagaceae 45
Asparagus **45**
  densiflorus **45**
  officinalis **45**
asparagus 6, **45**
  fern 45
asparagus, climbing 45
Asphodelaceae 152
Asplenium scolopendrium 25
Asteraceae 17, 133
Astrantia 25
Athyrium 25
Atropa belladonna 12, 33, **71**
Aubretia 27
Aucuba japonica **29**
autumn crocus 33, **62**

squill **55**
Azalea 114
azalea **114**
azalea, mock **150**

baby's breath 26
Ballota pseudodictamnus 25
baneberry **41**
basil 27
bay 17, **30**
bay, rose **115**
bean, coral **154**
    prayer **154**
beans, lucky 33, **154**
    paternoster **154**
bear's foot **51**
bedding lobelia 28
bee balm 26
belladonna **71**
    lily **61**
belladonna, Russian 33, **70**
Bellis perennis 26
Benjamin tree **135**
Berberidaceae 50
Berberis 18
    darwinii **29**
    thunbergii **29**
Bergenia 25
Beta vulgaris 27
bittersweet **75**
black bryony **47**
    cohosh **41**
    dogwood **90**
    locust **120**
    nightshade **72**
blood flower 17
blow **96**
blue taro **163**
bluebell **10**, **56**
blueberry 12, **32**
blushing philodendron **166**
bog arum **39**
bonnet, Granny's 27
borage 12
Boraginaceae 127, 128
Borago officinalis 12
Boston fern 17
bottle gourd **143**
Bracteantha 17
    bracteata 26
bramble 18
Brassaia actinophylla **137**
Briza maxima 25
Brompton stock 26
broom laburnum **118**
broom, purple 118
    Spanish **116**
Brugmansia 33, 68, **140**
    arborea **140**
    × *insignis* **140**
Brunfelsia **141**
    pauciflora 'Eximia' **141**
Bryonia cretica subsp. *dioica* 46
    dioica **46**
bryony, black **47**
    red **46**
    white **46**

buckeye **95**
buckthorn **90**
Buddleja 27
Bulbocodium 62
bunny ears 19, **20**, **153**
burning bush 14, **112**
bushman's poison **148**
busy Lizzie 28
butterfly bush 27

cabbage 27
cabbage, skunk **40**
Cactaceae **153**
cacti 19
cactus, redbird **151**
    slipper **151**
calabash **143**
Caladium **160**
    'Aaron' **160**
    'Pink Beauty' **160**
Calendula officinalis 27
calico bush **113**
California beauty **93**
    poppy 28
Calla hort. 38
Calla palustris **39**, 169
calla **38**, 39
    lily **38**
calla, florists' 38
Campanula medium 27
Campanulaceae **123**
Canna indica 26
Cannabaceae 96
Cannabis sativa **96**, 169
Canterbury bells 27
Cape tulip **63**
Caprifoliaceae 79, 88
Capsicum annuum **77**
cardinal flower **123**
Carissa acokanthera 148
Carolina jessamine **145**
carrot family 12
Caryophyllaceae 129
Cascabela thevetia **147**
Castanea sativa 95
castor bean plant **98**
castor oil plant **8**, 13, 30, 33, **98**
castor oil plant, false **30**, 91, 98
cat's tail **139**
celandine, greater **110**
Celastraceae 81
celeriac **107**
celery 6, 14, **107**
Celosia 28
Centaurea cyanus 28
    moschata 26
Cerasus 87
Cerbera thevetia **147**
chalice vine 33, **142**
Chamaecyparis nootkatensis 132
chard, ruby 27
    Swiss 27
chaste tree **97**
cheese plant **167**
Cheiranthus 26
Chelidonium majus **110**
chenille plant **139**

cherry laurel 23, **87**
cherry, Christmas **76**
　　Jerusalem **76**
　　winter **76**
chestnut, horse **95**
　　sweet 95
chicory 17
Chilean potato tree **74**
chilli pepper **77**
　　'Numex Twilight' **77**
Chimonanthus praecox 27
chincherinchee **54**
Chinese evergreens **155**
　　lantern **31**
Chionodoxa 28
chives 27
chocolate cosmos 26
Christmas cherry **76**
　　rose **51**
Chrysanthemum 17, **133**
　　'Misty Golden' and 'Misty Lemon' **133**
　　'Mona Lisa Pink' **133**
Cichorium 17
Cicuta virosa 33, **105**
climbing asparagus **45**
Clusiaceae 109
cobra-lily **36**
cockscomb 28
cocoyam **162**
Codiaeum 17
cohosh, black **41**
Colchicaceae 62, 144
Colchicum 33, **62**
　　bornmuelleri **62**
　　speciosum **62**
Coleus 17
Colocasia esculenta **162**
　　gigantea **162**
　　indica 161
　　macrorrhizos 161
comfrey 12, 22, **127**
Compositae 133
Conium maculatum 6, 33, **103**
conker tree **95**
Consolida 124
consolida **124**
Convallaria 33, **43**
　　majalis **43**
Convallariaceae 43, 44
Convolvulaceae 48
copperleaf **139**
coral bean **154**
　　bells 25
Coriaria 33, **83**
　　myrtifolia **83**
Coriariaceae 83
corncockle 13, **129**
cornflower 28
Cortaderia 18
Cosmos atrosanguineus 26
　　bipinnatus 28
cosmos, chocolate 26
Cotoneaster 12
　　horizontalis **29**
cow parsley **10**, 14, **102**
　　parsnip **106**
cowbane 33, **105**

cowbane, water **105**
crab's eyes **154**
cranesbill 28
crape or crepe jasmine **149**
Crataegus monogyna **30**
Crocosmia masonorum 28
Crocus vernus 28
crocus, autumn 33, **62**
croton, variegated 17
crown of thorns **151**
cuckoo pint 33, **35**
Cucurbita lagenaria 143
Cucurbitaceae 46, 143
cup of gold **142**
Cupressaceae 132
× ***Cupressocyparis leylandii*** **132**
Cupressus macrocarpa 132
Cymbidium 17
Cynara scolymus 17
cypress, Leyland **132**
Cypripedium 17
Cytisus laburnum 119
　　purpureus 118

daffodil 12, 17, **58**
Dahlia 17, 26
daisy family 17
daisy, pompom 26
danewort **88**
Daphne 33, **86**
　　laureola **86**
　　mezereum **86**
　　tangutica **86**
dasheen **162**
Datura 33, **68**, 140
　　'La Fleur Lilac' **68**
datura, tree 33, **140**
dead man's fingers **104**
deadly nightshade 12, 30, 33, **71**
death camas 33, **53**
Delphinium **124**
　　Belladonna Group **124**
Dendranthema 133
desert rose **150**
Devil's ivy **165**
Dianthus 26
　　barbatus 26
Dictamnus albus 14, **112**
Dieffenbachia 14, 20, 33, **156**
　　maculata 'Compacta' **156**
　　seguine **156**
Digitalis 9, 12, 22, 33, **126**
　　grandiflora 9, **126**
　　purpurea 9, **22**, **126**
dinner plate aralia **136**
Dioscorea communis **47**
Dioscoreaceae 47
dittany **112**
dittany, false 25
dogbane **131**
dogwood, black **90**
dope **96**
Dracunculus **37**
　　vulgaris **37**
dragon arum **37**
dropwort, hemlock water 12, 33, **104**
Duke of Argyll's tea plant **31**

dumbcane 14, 20, 33, **156**
dwale **71**
dwarf elder **88**

ears, bunny 19, **20, 153**
   elephant's 25, **160, 162**
   lamb's 25
   giant elephant's **161**
East Indian rosebay **149**
***Echium* 128**
   vulgare **128**
eddo **162**
elder **88**
elderberry **88**
elephant's ears 25, **160, 162**
elephant's ears, giant **161**
Endymion 56
English marigold 27
***Epipremnum* 165**
   aureum 164, **165**
   pictum 'Argyraeum' **165**
Eranthis hyemalis 125
Ericaceae 78, 113, 114
Ervatamia 149
Erysimum 26
Eschscholzia californica 27
***Euonymus* 81**
   europaeus **81**
***Euphorbia* 14, 20, 33**
   (Indoors) 151
   (Outdoors) 99
   characias **99**, 169
   griffithii 'Fireglow' **99**
   lathyris **99**
   milii **151**
   pulcherrima 20, 99, 151
   tithymaloides **151**
Euphorbiaceae 98, 99, 139, 151
evening primrose 26
   trumpet flower **145**

Fabaceae 116, 117, 118, 119, 120, 121, 122, 154
false acacia **120**
   castor oil plant **30**, 91, 98
   dittany 25
   hellebore 12, 33, **52**
   jalap **65**
   yellow jasmine **145**
Fatsia japonica **30**, 91, 98
fennel 27
ferns 17, 25
fern, asparagus **45**
   Boston 17
   foxtail **45**
   hardy shield 25
   hart's tongue 25
   ladder 17
   lady 25
   leather leaf 17
   sword 17
feverfew 17
***Ficus benjamina* 135**
     'Danielle' **135**
     'Starlight' **135**
   carica 14, **94**
fig 14, **94**

fig, weeping **135**
firethorn **19, 32**
flag, yellow **42**
flame lily 33, **144**
flamingo flower **158**
flannel bush **93**
florists' calla **38**
   chyrsanthemum **133**
fountain grass 25
four o'clock plant **65**
foxglove 9, 12, **22**, 33, **126**
foxtail fern **45**
Frangula alnus 90
Fremontia 93
***Fremontodendron* 18, 93**
   'California Glory' **93**
Fuchsia 26, **30**

garden rue 14, 33, **111**
***Gaultheria mucronata* 78**, 169
   procumbens 78
   section *Pernettya* 78
gayfeather 26
Gazania 28
Gelsemiaceae 145
***Gelsemium* 145**, 169
gentian, yellow 12
Gentiana lutea 12
Geranium 28
geranium 28
German primula 18, 33, **138**
giant elephant's ears **161**
   hogweed **106**
   silver mullein 25
   taro **161**
gladdon **42**
globe artichoke 17
***Gloriosa superba* 33, 144**
     'Lutea' **144**
     'Rothschildiana' **144**
glory lily 33, **144**
glory, morning **48**
goji berries **31**
golden chain **119**
   pothos **165**
   rain 33, **119**
   trumpet **146**
gooseberry 18
gourd, bottle **143**
Granny's bonnet 27
   pincushion 25
grape hyacinth 28
grass **96**
grass, fountain 25
   pampas 18
   quaking 25
grasses 18, 25
greater celandine **110**
Guttiferae 109
Gypsophila 26

hardy geranium 28
   shield fern 25
hare's tail 25
hart's tongue fern 25
hash **96**
hawthorn **30**, 80

heart leaf **166**
Hedera 17, **91**
   helix **16**
Helenium 17
Helianthus annuus 17, 28
Helichrysum 17
   bracteatum 26
heliotrope 26
Heliotropium arborescens 26
hellebore 14, **51**
hellebore, false 12, 33, **52**
Helleborus 14, **51**
   foetidus **51**
helmet flower **125**
hemlock 6, 33, **103**
   water dropwort 12, 33, **104**
hemlock, water 33, **105**
hemp **96**
hemp, Indian **131**
henbane 33, **67**
Heracleum 14, 33, **106**
   mantegazzianum **106**
   sphondyllium **106**
herb Christopher **41**
   of grace **111**
Heuchera 25
Hippeastrum **60**, 61
Hippocastanaceae 95
hogweed 14, 33, **106**
hogweed, giant **106**
holly 18, **30**
Homeria 63
honeysuckle 27, **31**
honeysuckle, shrubby 27
hops 18
horse chestnut **95**
Hosta 25
Humulus lupulus 18
hyacinth 14, 17, **57**
hyacinth, grape 28
Hyacinthaceae 54, 55, 56, 57
Hyacinthoides 55, **56**
   hispanica **56**
   non-scripta **10**, **56**
Hyacinthus 14, 17, **57**
Hydrangea 17
Hyoscyamus 33, **67**
   niger **67**
Hypericum *androsaemum* **30**, 71
   perforatum 109
   × *inodorum* 'Magical Red' **30**
hyssop, ornamental 26

Ilex 18, **30**
impala lily **150**
Impatiens walleriana 27
Indian hemp **131**
   kale **163**
   shot plant 26
inkberry 33, **84**
***Ipomoea purpurea* 48**
     'Star of Yelta' **48**
   rubrocaerulea 48
   ***tricolor* 48**
     'Heavenly Blue' **48**
   violacea hort. 48
Iridaceae 42, 63

Iris 28, **42**
   foetidissima **42**, 169
   reticulata 28
Italian lords-and-ladies **35**
ivy **16**, 17, 30, **35**, **91**
ivy, Devil's **165**
   poison 15, 33, **92**

Jack-in-the-pulpit **36**
Jackman's blue **111**
Jacob's coat **139**
Jamestown weed **68**
jasmine, crape or crepe **149**
   false yellow **145**
jequerity or jequirity seeds **154**
Jersey lily **61**
Jerusalem cherry **76**
jimsonweed 33, **68**
jonquil **58**

Kalmia **113**
   angustifolia **113**
   latifolia **113**
     'Pink Charm' **113**
kangaroo apple **74**
   paws 17
keck **102**, **106**
kowhai 33, **117**
kudu lily **150**

Labiatae 97
+*Laburnocytisus* 'Adamii' 33, **118**
Laburnum **5**, 11, **12**, 33, **119**, 169
   anagyroides 118, **119**
   × *watereri* 'Vossii' 23
laburnum, Adam's 33, **118**
   broom **118**
ladder fern 17
ladies, lords-and- 33, **35**
   naked 33, **62**
lady fern 25
   of the night **141**
Lagenaria **143**
   siceraria **143**, 169
   vulgaris 143
Lagurus ovatus 25
lamb's ears 25
Lamiaceae 97
Lantana 33, **85**
   camara **85**
   'Lucky Lemon Cream' **85**
lantern, Chinese **31**
larkspur **124**
   'Stock Flowered Mixed' **124**
Lathyrus odoratus 26
laurel 17, **31**
laurel, American **113**
   cherry 23, **87**
   Portugal **87**
   rose **115**
   spotted **29**
   spurge 33, **86**
   true **31**
Laurocerasus 87
Laurus nobilis 17, **31**, 87
Lavandula 26
lavender 26

leather leaf fern 17
Leguminosae 116, 117, 118, 119, 120, 121, 122, 154
Lent lily **58**
leopard lily 33, **156**
lettuce 27
Leyland cypress **132**
Liatris spicata 26
Ligustrum 23, **89**
    ovalifolium **89**
    vulgare **89**
lilac 27
Liliaceae 59
lily 20
lily of the Incas **64**
lily-of-the-valley 33, **43**
lily, arum **38**
    belladonna **61**
    calla **38**
    cobra- **36**
    flame 33, **144**
    glory 33, **144**
    impala **150**
    Jersey **61**
    kudu **150**
    Lent **58**
    leopard 33, **156**
    peace 20, **157**
    Peruvian **16**, 17, **64**
    plantain 25
Limnanthes douglasii 28
Limonium sinuatum 26
Lobelia **123**
    erinus 28, 123
    siphilitica **123**
    tupa **123**
    'Scarlet Fan' **123**
lobelia, bedding 28
Lonicera fragrantissima 27
    periclymenum 27, **31**
    pileata **31**
lords-and-ladies 33, **35**
love in a mist 26
love lies bleeding 25
lucky beans 33, **154**
lupin **7**, **122**
Lupinus **7**, **122**
    polyphyllus **122**
    'Tom Tom the Piper's Son' **122**
Lychnis githago 129
Lycium barbarum 31
    chinense **31**
Lysichiton **40**
    americanus **40**, 169
    camtschatensis **40**, 169

Mahonia japonica **31**
Malvaceae 93
Mandragora 33, **69**
    officinarum **69**
mandrake 33, **69**
marigold, English or pot 27
marijuana **96**
marjoram 27
marvel of Peru **65**
Matthiola incana 26
May apple **50**

meadow saffron **62**
Melianthaceae 52, 53
Merendera 62
mezereon 33, **86**
mignonette 26
mimosa 120
Mimulus 28
mint 27
Mirabilis **65**
    jalapa **65**
mistletoe **80**
mock azalea **150**
Monarda didyma 26
money plant **165**
monkey flower 28
monkshood 20, 33, **125**
Monstera deliciosa **167**
montbretia 28
Moraceae 94, 135
Moraea *collina* **63**
    elegans **63**
    longistyla **63**
    section ***Homeria*** **63**
morning glory **48**
mountain ash **32**
mullein, giant silver 25
Muscari 28
    latifolium 28

naked ladies 33, **62**
Narcissus 12, 17, **58**, 169
nasturtium 11, **12**, 28
necklace pod **117**
Nemesia strumosa 28
Nephrolepis exaltata 17
Nerium oleander 33, **115**, 147
nettle **5**, 14, **100**
Nicotiana **7**, 66
    sylvestris **66**
    sanderae 'Malibu Blue' and 'Malibu Lime' **66**
Nigella damascena 26
nightshade **74**
nightshade, black **72**
    deadly 12, 30, 33, **71**
    woody **13**, 33, **75**
Nyctaginaceae 65

Oenanthe *aquatica* 104
    crocata 12, 33, **104**
    javanica 104
    phellandrium 104
    pimpinelloides 104
    sarmentosa 104
Oenothera biennis 26
oil of wintergreen 78
Oleaceae 89
oleander 33, **115**, 147
oleander, yellow 33, **147**
onion 17
onions, ornamental 25
opium poppy 9, **130**
Opuntia 19
    microdasys 19, **20**, **153**
    f. *alba* **153**
orchids 17
ornamental hyssop 26

onions 25
pepper **77**
sage 26
***Ornithogalum* 54**
   longibracteatum 54
   saundersiae 54
   umbellatum 54
Osteospermum ecklonis 28
Oswego tea 26

painter's palette 20, **158**
palma Christi **98**
pampas grass 18
pansy 28
Papaver somniferum 9, **130**
Papaveraceae 110, 130
Paphiopedilum 17
parsley, cow **10**, 14, **102**
parsnip 6, 14, **108**
parsnip, cow **106**
pasqueflower 17
***Passiflora* 49**
   caerulea **49**, 169
Passifloraceae 49
passion flower or fruit **49**
Pastinaca sativa 14, **108**, 169
paternoster beans **154**
pea 27
pea, rosary 33, **154**
   sweet 26
peace lily 20, **157**
peach 11, 87
Pedilanthus 151
   tithymaloides 151
Pelargonium 28
Pennisetum alopecuroides 25
Penstemon 28
pepper, chilli **77**
   ornamental **77**
perforate St John's-wort **109**
Pernettya 78
pernettya **78**
Peru, marvel of **65**
Peruvian lily **16**, 17, **64**
Philodendron 8, 20, **166**
   bipinnatifidum **166**
   erubescens **166**
Phlox paniculata 26
Physalis alkekengi **31**
Phytolacca 33, **84**
   americana 12, **84**
   clavigera **84**
Phytolaccaceae 84
pigeonberry **84**
pigtail plant **158**
pincushion, Granny's 25
pinks 26
pinwheel flower **149**
Plantaginaceae 126
plantain lily 25
plum 87
poached egg plant 28
***Podophyllum* 50**
   hexandrum **50**
   peltatum **50**
poinsettia 20, 99, 151
poison bush **148**

ivy 15, 33, **92**
primula 33, **138**
poison, bushman's **148**
pokeroot **84**
pokeweed 12, 33, **84**
polyanthus 28
Polygonaceae 101
***Polygonatum* 44**
   latifolium **44**
   maximowiczii **44**
   verticillatum **44**
***Polyscias* 136**
   cumingiana **136**
   guilfoylei **136**
   'Filicifolia' **136**
Polystichum aculeatum 25
pompom daisy 26
ponytails 25
poppy, California 28
   opium 9, **130**
Portugal laurel **87**
pot **96**
   marigold 27
potato 6, **73**
   vine **74**
potato tree, Chilean **74**
Potentilla atrosanguinea 25
Pothos aureus 165
pothos, golden **165**
prayer bean **154**
prickly heath **78**
primrose, evening 26
Primula obconica 18, 33, **138**
   Polyanthus Group 28
primula, German 18, 33, **138**
   poison 33, **138**
Primulaceae 138
privet 23, **89**
Prunus laurocerasus 11, 23, **87**
   lusitanica 11, **87**
purple broom 118
Pyracantha 12, 18, **19**, **32**

quaking grass 25
Queen Anne's lace **102**

Ranunculaceae 41, 51, 124, 125
Raphidophora 164
raspberry 18
red bryony **46**
redbird cactus **151**
Reseda odorata 26
Rhamnaceae 90
***Rhamnus* 90**
   cathartica 90
   frangula **90**
   purshiana 90
   tinctoria **90**
***Rhaphidophora* 164**
   aurea 164, 165
   latevaginata **164**
   petrieana **164**
Rheum rhabarbarum 101
   × *cultorum* 101
   × ***hybridum* 101**
Rhododendron 5, 11, **114**
   ponticum **114**

× *loderi* **114**
rhubarb 6, **101**
Rhus radicans 15, 33, **92**
   succedanea 92
   typhina 92
   verniciflua 92
Ricinus communis 8, 13, 33, **98**, 169
Robinia 24
   pseudoacacia **120**
Rosa 27
Rosaceae 87
rosary pea 33, **154**
rose 18, 27
   bay **115**
   laurel **115**
rose, Christmas **51**
   desert **150**
rosebay, East Indian **149**
rosemary 27
rowan **32**
rubber plant, small-leaved **135**
ruby chard 27
rue **111**
   garden 14, 33, **111**
Ruscaceae 43, 44
Russian belladonna 33, **70**
   comfrey **127**
Ruta 14, 33, **111**
   graveolens **111**
Rutaceae 111, 112

sabi star **150**
saffron, meadow **62**
sage 27
sage, silver 25
   ornamental 26
Salvia 26
   argentea 25
Sambucus **88**
   adnata **88**
   nigra **88**
Santalaceae 80
Sapindaceae 95
Scabiosa 28
   atropurpurea 26
Schefflera **137**
   'Compacta' **137**
   'Trinette' **137**
Scilla **55**
   campanulata 56
   italica 56
   non-scripta 56
   nutans 56
   peruviana **55**
   siberica **55**
Scindapsus aureus 165
Scopolia 33, **70**
   carniolica **70**
sedges 18
shingle plant **164**
shrub verbena 33, **85**
shrubby honeysuckle 27
silver mullein, giant 25
   sage 25
skunk cabbage **40**
slipper cactus **151**
   spurge **151**

small-leaved rubber plant **135**
snail-flower **36**
snapdragon 25
sneezeweed 17
snowberry **79**
Solanaceae 66, 67, 68, 69, 70, 71, 72, 73, 74, 75, 76, 77, 140, 141, 142
Solandra 33, **142**
   maxima **142**, 169
Solanum **(other cultivated ornamental species) 74**
   aviculare 74
   capsicastrum 76
   crispum 'Glasnevin' **74**
   diflorum 76
   dulcamara **13**, 33, **75**
   jasminoides 'Album' 74
   laciniatum **74**
   laxum 'Album' **74**
   nigrum **72**
   pseudocapsicum **76**
   rantonettii 74
   tuberosum **73**
Solomon's seal **44**
Sophora 33, **117**
   microphylla 'Sun King' **117**
Sorbus **32**
   aucuparia **32**
   hupehensis **32**
Spanish bluebell **56**
   broom **116**
Spartium junceum **116**
Spathiphyllum 20, **157**
   blandum **157**
spinach 12
spindle tree **81**
spotted laurel **29**
spring squill **55**
spurge 14, 33, **99**
   laurel 33, **86**
spurge, slipper **151**
squill 55
squill, autumn **55**
   spring **55**
St John's-wort **109**
St Martin's flower **64**
Stachys byzantina 25
   lanata 25
star-of-Bethlehem **54**
statice 26
Sterculiaceae 93
stinging nettle 14, **100**
stinkweed **68**
Stipa tenuissima 25
stock, Brompton 26
straw flower 17, 26
strawberry 27
sumach 92
sunflower 17, 28
sweet chestnut 95
   pea 26
   sultan 26
   William 26
sweetheart plant **166**
Swiss chard 27
   cheese plant **167**
sword fern 17

Symphoricarpos **79**
   albus **79**
   × *chenaultii* 'Hancock' **79**
Symphytum 12, 22, **127**
   officinale **127**, 169
   orientale **127**
   × *uplandicum* **127**
Syngonium **159**
   auritum **159**
Syringa 27

Tabernaemontana **149**
   divaricata **149**
tailflower **158**
Tamus communis 47
Tanacetum 17
tansy 17
taro **162**
taro, blue **163**
   giant **161**
Taxaceae 82
Taxus 21, 23, 33, **82**
   baccata 'Fastigiata' **82**
Thevetia 33, 115, **147**
   peruviana **147**, 169
thornapple 33, **68**
thyme 27
Thymelaeaceae 86
tobacco **7**, **66**
Toxicodendron 92
tree datura 33, **140**
Tropaeolum majus **12**, 28
trumpet flower, evening **145**
trumpet, golden **146**
trumpets, angels' 33, **140**
tulip 17, **59**
tulip, Cape **63**
Tulipa 17, **59**
tutsan **30**, 71
tutu 33, **83**

Umbelliferae 102, 103, 104, 105, 106, 107, 108
umbrella plant or tree **137**
Urtica dioica 5, 14, **100**
Urticaceae 100

Vaccinium corymbosum **32**
variegated croton 17
Veratrum 12, 33, **52**
   album **52**, 169
   nigrum **52**
Verbascum bombyciferum 25
Verbena bonariensis 28
   speciosa 28
verbena, shrub 33, **85**
Verbenaceae 85, 97
Veronica 28
Viburnum 27
vine, arrowhead **159**
   chalice 33, **142**

potato **74**
Viola odorata 26
   × *wittrockiana* 28
violet 26
viper's bugloss **128**
Viscaceae 80
Viscum album 80
Vitex 97
   agnus-castus var. *latifolia* **97**
     f. *alba* **97**

wallflower 26
water arum **39**
   cowbane **105**
   hemlock 33, **105**
water dropwort, hemlock 12, 33, **104**
weed **96**
weeping fig **135**
white bryony **46**
   comfrey **127**
   wild celery **107**
   parsnip **108**
wings, angel's **160**
winter aconite 125
   cherry **76**
wintergreen **78**
wintergreen, oil of 78
wintersweet 27, **148**
Wistaria 121
Wisteria 121
   floribunda 121
     'Alba' **121**
wolf's bane 33, **125**
wonder flower **54**
woody nightshade **13**, 33, **75**

Xanthorrhoeaceae 152
Xanthosoma **163**
   sagitifolium **163**
   violaceum **163**

yarrow 17
yautia **163**
yellow flag **42**
   foxglove **9**
   gentian 12
   oleander 33, **147**
yellow jasmine, false **145**
yesterday, today and tomorrow **141**
yew 21, 23, 33, **82**
Yucca 18

Zantedeschia **38**, 39
   aethiopica **38**
   elliottiana **38**
Zigadenus 33, **53**
   venenosus **53**
Zinnia 26
Zygadenus 53